Gettin Into Geometry

Developed and Published
by
AIMS Education Foundation

This book contains materials developed by the AIMS Education Foundation. **AIMS** (**A**ctivities **I**ntegrating **M**athematics and **S**cience) began in 1981 with a grant from the National Science Foundation. The non-profit AIMS Education Foundation publishes hands-on instructional materials that build conceptual understanding. The foundation also sponsors a national program of professional development through which educators may gain expertise in teaching math and science.

Copyright © 2010 by the AIMS Education Foundation

All rights reserved. No part of this book or associated digital media may be reproduced or transmitted in any form or by any means—graphic, electronic, or mechanical, including information storage/retrieval systems—except as noted below.

- A person or school purchasing this AIMS publication is hereby granted permission to make up to 200 copies of any portion of it, provided these copies will be used for educational purposes and only at one school site. For a workshop or conference session, presenters may make one copy of a purchased activity for each participant, with a limit of five activities per workshop or conference session.
- Workshop or conference presenters may make one copy of a purchased activity for each participant, with a limit of five activities per workshop or conference session.
- All copies must bear the AIMS Education Foundation copyright information.

AIMS users may purchase unlimited duplication rights for making more than 200 copies, for use at more than one school site. Contact us or visit the AIMS website for complete details.

New Copyright Policy!

A person purchasing this AIMS publication is hereby granted permission to make unlimited copies of any portion of it (or the files on the accompanying disc), provided these copies will be used only in his or her own classroom. Sharing the materials or making copies for additional classrooms or schools or for other individuals is a violation of AIMS copyright. Please visit **www.aimsedu.org** for further details.

AIMS Education Foundation
P.O. Box 8120, Fresno, CA 93747-8120 • 888.733.2467 • aimsedu.org

ISBN 978-1-60519-040-2

Printed in the United States of America

Getting Into Geometry

Table of Contents

Chinese Proverb 4

2-D Shapes
Shape Sense..................... 5
Making Models 9
Shape Takers.................. 15
Stepping Out Shapes....... 21
Art in the Air................... 27
Sid the Snake 29
Sandy Shapes 35
Rough Enough Shapes 37
Shapes Line Up 45
Fabulous Folds 51
Find the Fit..................... 53
Shapes on the Shelf........ 57
Goin' on a Shape Hunt.... 61
Shapes on the Bus.......... 67
Shape World 75
Animals Shape Up........... 77

Composing and Decomposing Shapes
Puzzle Makers................. 81
Piece-by-Piece Pictures .. 87
Picture Perfect 95
The Big Picture 103

Symmetry
Shape Symmetry............ 109
Match Play.................... 117
Squishy Symmetry 121
The Art of Symmetry..... 123

3-D Solids
3-D Explorations 129
The Shape-Up Song 133
Celebrating Solids......... 135
Solids on the Slide 143
Make a Match 149
Behind the Curtain....... 151
All About Town 153
Same Shapes 155
Rootin' Tootin' Relay..... 161
Solid Combinations 165

Spatial Relationships: Location/Position
Characters in Position... 171
Shapes on Location 175
Putting Shapes in Their Place 179
The Tinker's Toy Store... 185
Shape Snapshots.......... 193
Getting Around Geoville 195

Playful Practice
Spin and Win................ 201
What's in My Pocket? ... 205
Boat Builders............... 213
Twisting to Shapes....... 221
Eyes-on Geometry 227
Symmetry Pair-O 229
Matchmakers................ 233
Concentrating on Shapes and Solids.............. 239
Who Has? Shapes and Solids......... 243

Literature for Geometry 255

I Hear and
I Forget,

I See and
I Remember,

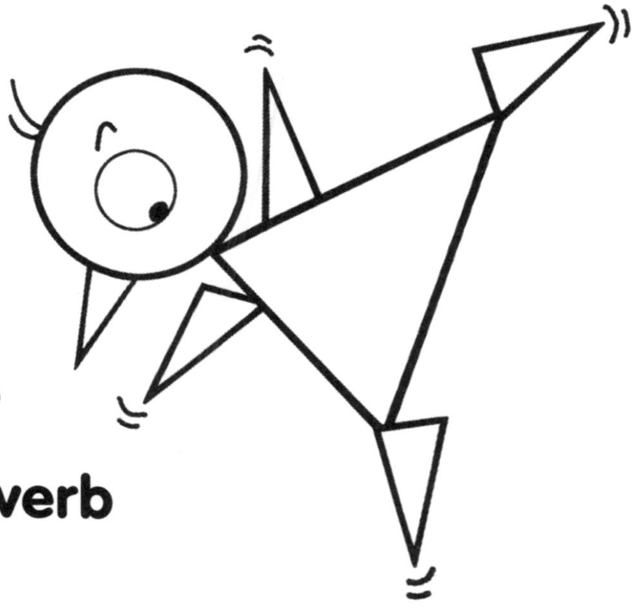

I Do and
I Understand.

- Chinese Proverb

Shape Sense

Topic
2-D shapes

Key Question
How can we describe the four basic shapes?

Learning Goal
Students will name and describe squares, circles, rectangles, and triangles.

Guiding Documents
Project 2061 Benchmarks
- *Numbers and shapes can be used to tell about things.*
- *Shapes such as circles, squares, and triangles can be used to describe many things that can be seen.*

NCTM Standards 2000*
- *Recognize, name, build, draw, compare, and sort two- and three-dimensional shapes*
- *Describe attributes and parts of two- and three-dimensional shapes*
- *Recognize and represent shapes from different perspectives*

Math
Geometry
    2-D shapes
        characteristics
        identification

Integrated Processes
Observing
Comparing and contrasting
Communicating

Materials
For each student:
    set of shapes (see *Management 1*)

For the class:
    chalkboard/whiteboard
    sentence strips (see *Management 3*)
    colored markers

Background Information
This should be one of the first lessons on shapes for young children. It will focus on the four basic geometric shapes—the circle, square, rectangle, and triangle. Students will learn the names of the shapes, as well as their characteristics.

Management
1. Copy and cut out a set of shapes for each student. Use multiple colors. If available, a die cut machine can be used to make the shapes.
2. For *Part Two* of the lesson, remove a few of each shape so that there will not be equal numbers of shapes.
3. In *Part Two,* the sentence strips will be cut to lengths appropriate for graphing labels.

Procedure
Part One
1. Display the four basic shapes—square, circle, rectangle, and triangle. Ask the students if they know the names of any of the shapes displayed.
2. Tell the class that these are shapes that they will see every day and that it is important for us to recognize them. Explain that they will be learning the names of the shapes, counting edges and corners, and sorting the shapes by their names and characteristics.
3. Give each student a set of shapes—square, circle, triangle, and rectangle. Place the circle on the board and write the word *circle* beneath it. Tell the class that they will first learn about circles. Ask the students to find the circle in their set. Question the students about the number of edges and corners it has. Ask them to try rolling the circle. Have the students describe the shape in their own words. Record the student responses under the name of the shape.
4. Repeat this procedure with the triangle, square, and rectangle.
5. When all shapes have been addressed, describe a shape and ask students to hold up the shape you are describing. For example, which shape has no straight edges? What is this shape's name?
6. Collect the shapes and store them for *Part Two* of the lesson. End with a discussion about where they might see the shapes that they learned about today.

GETTING INTO GEOMETRY   © 2010 AIMS Education Foundation

Part Two
1. Review the shape names and characteristics from *Part One*.
2. Gather the students into a circle. Place the pile of shapes in the center of the circle.
3. Invite a student to sort the shapes by color. Allow other students to help construct a concrete graph with the shapes. Using the markers and sentence strips, label the categories and give the graph a title. Ask various questions about the graph such as which color they have more of …less of …how many more ___ they have than ____, etc.
4. Invite another group of students to sort the shapes by the number of straight edges they have. [0, 3, 4] Label the concrete graph and discuss the results.
5. Repeat the sorting process using characteristics of the shapes such as number of corners, curved edges/not curved edges, etc.
6. End with time of sharing what they know about the shapes and ways that they were able to sort them.

Connecting Learning
1. What does a triangle look like? …square? …rectangle? …circle?
2. For which shape is it easiest for you to remember its name? Why do you think that is?
3. Which shape has no straight edges? …three straight edges? …four straight edges?
4. Describe a square. …circle. …rectangle. …triangle.
5. What are some ways that shapes can be sorted?
6. What two shapes are most alike? [the square and the rectangle] How are they alike? [They each have four straight edges and four corners.] How are they different? [The square has four equal edges, and the rectangle has opposite edges that are equal—two short and two long.]

* Reprinted with permission from *Principles and Standards for School Mathematics,* 2000 by the National Council of Teachers of Mathematics. All rights reserved.

Shape Sense

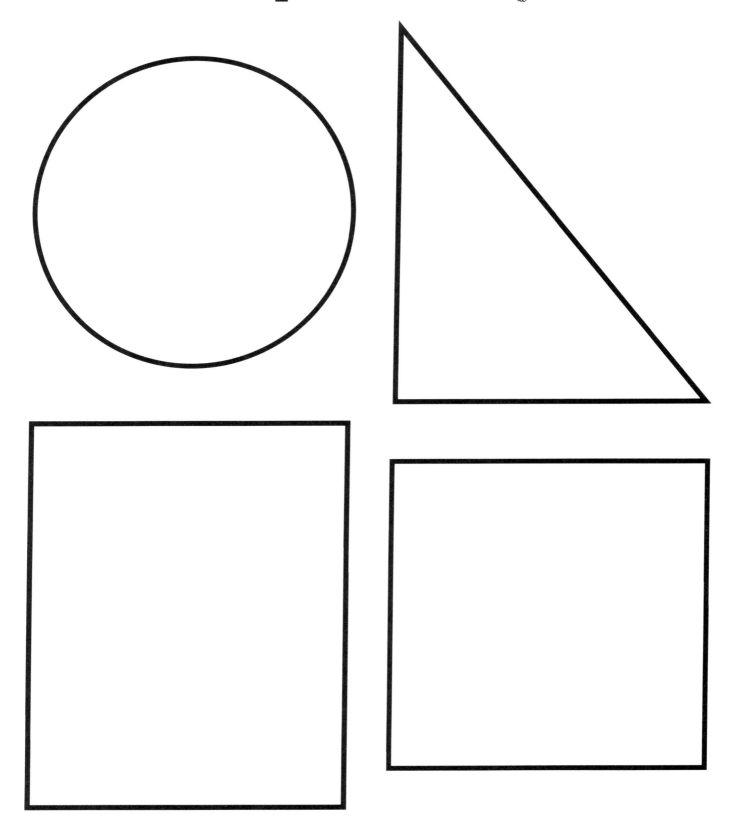

Making Models

Topic
2-D shapes

Key Question
How can we model the four basic shapes?

Learning Goal
Students will name, describe, and build models of the circle, triangle, square, and rectangle.

Guiding Documents
Project 2061 Benchmarks
- *Numbers and shapes can be used to tell about things.*
- *Shapes such as circles, squares, and triangles can be used to describe many things that can be seen.*

NCTM Standards 2000*
- *Recognize, name, build, draw, compare, and sort two- and three-dimensional shapes*
- *Describe attributes and parts of two- and three-dimensional shapes*
- *Recognize and represent shapes from different perspectives*

Math
Geometry
    2-D shapes
        modeling

Integrated Processes
Observing
Comparing and contrasting
Communicating

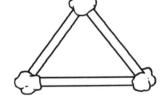

Materials
Straws (see *Management 2*)
Play-clay (see *Management 3*)
Yarn
Shape cards
Zipper-type plastic bag

Background Information
    Our world is filled with a variety of shapes. Other than circles and ovals, most regular shapes, are characterized by the number of straight edges and corners. Students will use straws to make rectangles, squares, and triangles. When they attempt to make a circle, they will discover that there are no straight edges on a circle, which means the straws are of no use. Instead, students will use yarn to make their circles. This activity will strengthen their knowledge of the characteristics of the shapes as well as their spatial sense.

Management
1. Prior to teaching this lesson, copy the circle, square, triangle, and rectangle onto card stock and laminate for extended use.
2. Prepare a set of straws for each student and place them in a zipper-type plastic bag. The straws should not be the bendable type. Each student will need a bag that includes nine long straws and seven short straws.
3. If available, use the soft clay that is similar to Play Dough®. Each student will need clay that is the approximate size of a table tennis ball.
4. Each student will need one 18-inch piece of yarn.

Procedure
1. Display the shape cards. Ask a student volunteer to come to the front of the class and point to the triangle.
2. Have the students look carefully at the triangle. Show them the straws and clay. Ask students if they can make the shape that they see using the straws as edges and the clay as corners. Demonstrate that they need a small bit of clay—less than the size of a small marble—to hold two straws together.
3. Repeat the procedure with the square and rectangle. Be sure to discuss differences and similarities between the two shapes.
4. Point to the circle and ask the students to name the shape. Ask them if they can make a circle with the straws and clay. Discuss the fact that it is not possible to make a shape that has no straight sides with straws. Brainstorm possible ways and materials to use to make a circle. [yarn, string, etc.]
5. Give each student a piece of yarn and ask them to make a circle.
6. End with a discussion about the shapes, their names, and characteristics of the shapes.

Connecting Learning
1. Could you make all four of the shapes? Explain.
2. How are the square and circle alike? …different?
3. What shape(s) could you make with three straws? …four straws?
4. How is the triangle different than the other shapes?
5. How are the squares and rectangles the same? How are they different?

* Reprinted with permission from *Principles and Standards for School Mathematics*, 2000 by the National Council of Teachers of Mathematics. All rights reserved.

Making Models

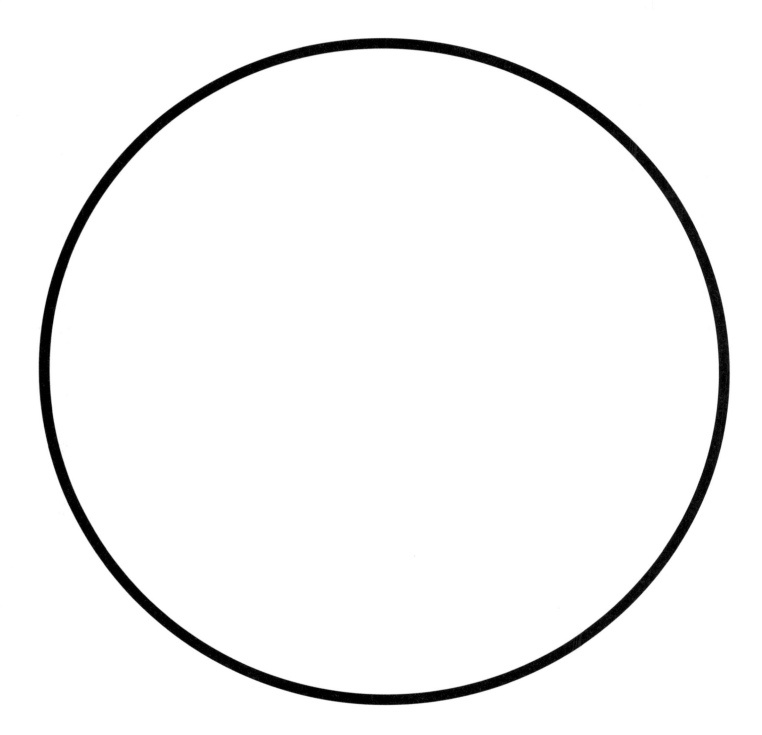

GETTING INTO GEOMETRY 10 © 2010 AIMS Education Foundation

Making Models

Making Models

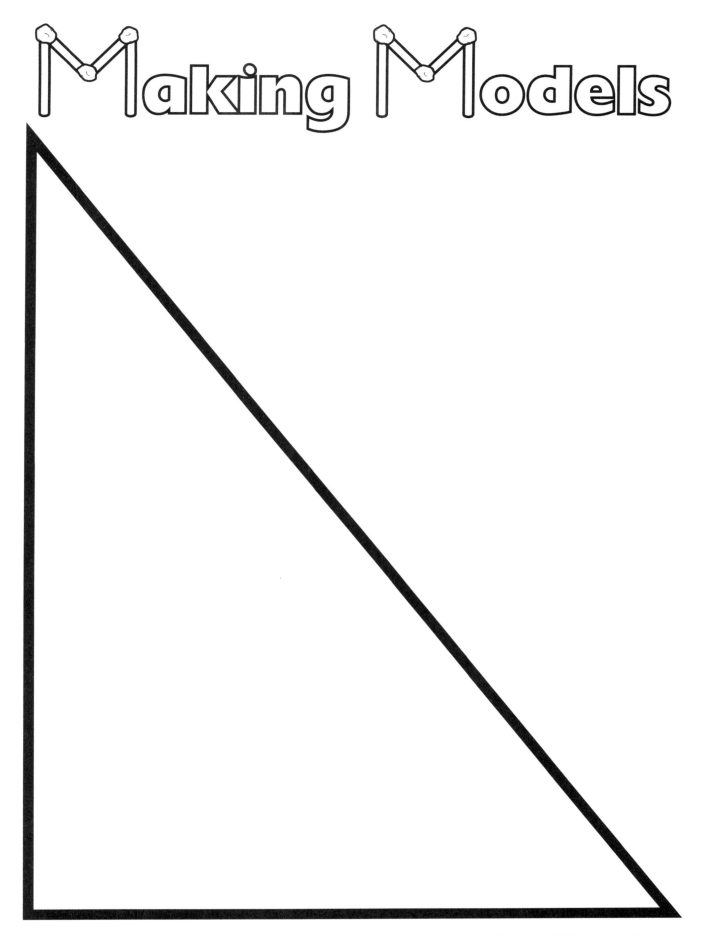

GETTING INTO GEOMETRY 13 © 2010 AIMS Education Foundation

Shape Takers

Topic
2-D shapes

Key Question
What shapes do we have and how are our shapes alike and different?

Learning Goals
Students will:
- describe how shapes of the same name are alike and different;
- sort and match shapes of different sizes and orientations; and
- sort shapes according to attributes such as straight edges, number of edges, and number of corners.

Guiding Documents
Project 2061 Benchmarks
- *Numbers and shapes can be used to tell about things.*
- *Shapes such as circles, squares, and triangles can be used to describe many things that can be seen.*

NCTM Standards 2000*
- *Recognize, name, build, draw, compare, and sort two- and three-dimensional shapes*
- *Describe attributes and parts of two- and three-dimensional shapes*
- *Recognize and represent shapes from different perspectives*

Math
Geometry
    2-D shapes
        characteristics
        identification

Integrated Processes
Observing
Recording data
Comparing
Communicating

Materials
For each student:
    one shape card (see *Management 1*)

For the class:
    class grid (see *Management 2*)
    chart paper
    lunch-size paper bag
    graph labels

Background Information
As young learners begin to explore two-dimensional shapes, one of the first things they must learn is what characteristics define a shape. For example, rectangles have four edges and four corners, but some are long and skinny, others are short and fat, etc. Viewing shapes of different sizes, in different orientations, with edges of different lengths, etc., is very important if students are to understand the attributes of each category. For example, children too often experience only one representation of a triangle—the equilateral triangle. They need to be exposed to right triangles, isosceles triangles, and scalene triangles as well. (Note, the naming of the triangles is *not* recommended at this time.)

The following is a list of attributes of the geometric shapes covered in this lesson. The list provides true statements that young learners need to consider and observe. Please allow the students to use their own words for defining and describing these attributes.
- All quadrilaterals (rectangle and square in this lesson) have four edges and four corners.
- All triangles have three edges and three corners.
- All edges on both quadrilaterals and triangles are straight.
- Not all shapes have the same length edges.
- Circles have no corners and no straight lines.
- Turning a shape only changes the orientation, not the attributes of the shape.

Management
1. On white card stock, duplicate and laminate (for extended use) one shape card for each student. Place the shape cards in a paper bag.
2. Prepare a class floor grid. Mark off four columns. Duplicate, cut apart, and laminate (for reuse) the *Graphing Labels*.
3. It is suggested that this activity be used as a reinforcement of the naming and classification of two-dimensional geometric shapes.

Procedure
Part One
1. Show the students the prepared shape cards. Discuss the names of the shapes: circle, square, rectangle, triangle.
2. Hand around the bag containing the shape cards, directing students to each draw out one shape. Ask them to state the geometric names of the shapes they are holding, and have them describe their shapes by naming the number of edges and corners.

GETTING INTO GEOMETRY   © 2010 AIMS Education Foundation

3. Ask the students to gather in groups according to the kinds of shapes on their cards.
4. Tell them to compare and contrast the shapes in their groups by observing the sizes, orientations on the cards, and numbers of edges and corners.
5. Invite the students to share statements that will be true for all the triangles. ...all the squares, etc. Record these statements on chart paper.

Part Two
1. Have the students sit in a circle on the floor with each student holding one shape card.
2. Using the following type of directions, play a listening game with the students. "If you are holding a triangle, stand up. If you are holding a rectangle, turn around. ...square, sit down." Continue calling on all the shapes being used in the game.
3. Direct the students to find someone who has a shape card with a shape that is different from the one they are holding. Ask the students to discuss with this classmate how their shapes are different. Tell them to trade shape cards.
4. Looking at their new shape cards, have each student describe the new shape and tell its geometric name.
5. Repeat the listening game several times.
6. To end the game and to collect the shape cards, direct the students to use the class floor grid to sort and display their shape cards according to the grid labels.
7. Discuss the grid with the students using the *Connecting Learning* questions that follow.

Connecting Learning
1. Show me a triangle. ...a square. ...a circle. ...a rectangle.
2. Choose one of the shapes and describe it to the rest of the class. Let the class guess which shape you are describing.
3. Tell the class something that is true about all triangles (rectangles, squares, and circles).
4. Show me two different triangles. ...rectangles. ...squares. ...circles. How are they alike and how are they different? Why are they both triangles? ...rectangles? ...etc.?
5. Using our class grid, which shape was found the most in our set of shape cards? ...the least?
6. Do all shapes of the same geometric name look exactly alike? Explain.

Extensions
1. Direct students to find other examples of these shapes in the room, outside, etc. Ask them to describe how they are alike and different from the shapes on the cards.
2. Ask the students to arrange each shape in order from smallest to largest.
3. Direct the students to form patterns, such as ABAB, using the shape cards. Encourage the students to use shape, orientation, size, etc., as part of the pattern rules.
4. Direct the students to record their shapes by making a shape rubbing. Demonstrate how to do a rubbing by cutting out a shape, placing it under a piece of paper, and rubbing the side of a crayon back and forth across the top surface of the paper. Urge them to change the position of their shapes and do other rubbings.

* Reprinted with permission from *Principles and Standards for School Mathematics,* 2000 by the National Council of Teachers of Mathematics. All rights reserved.

 Graphing Labels

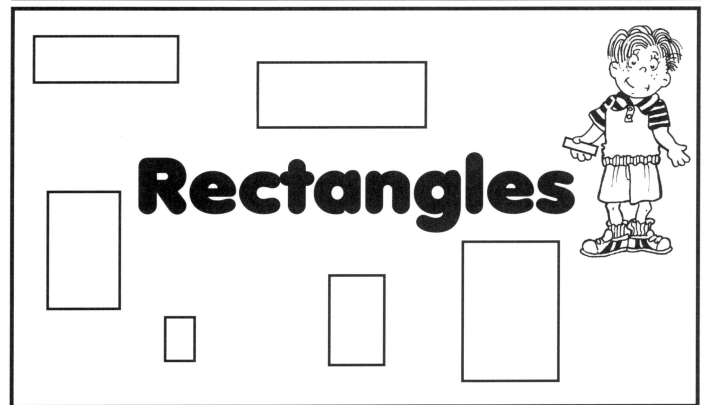

Graphing Labels

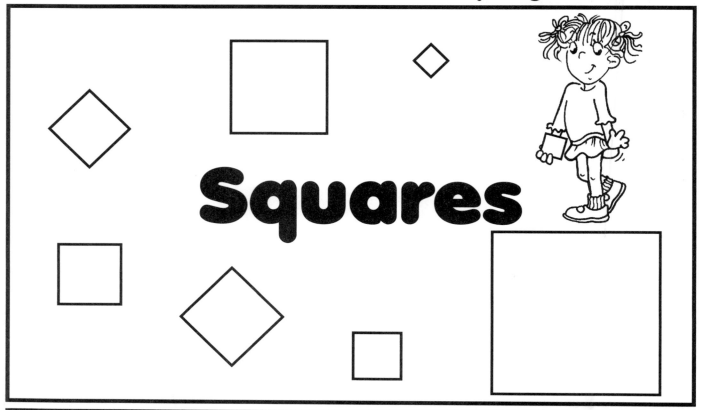

Squares

Triangles

GETTING INTO GEOMETRY © 2010 AIMS Education Foundation

Shape Takers   Shape Cards

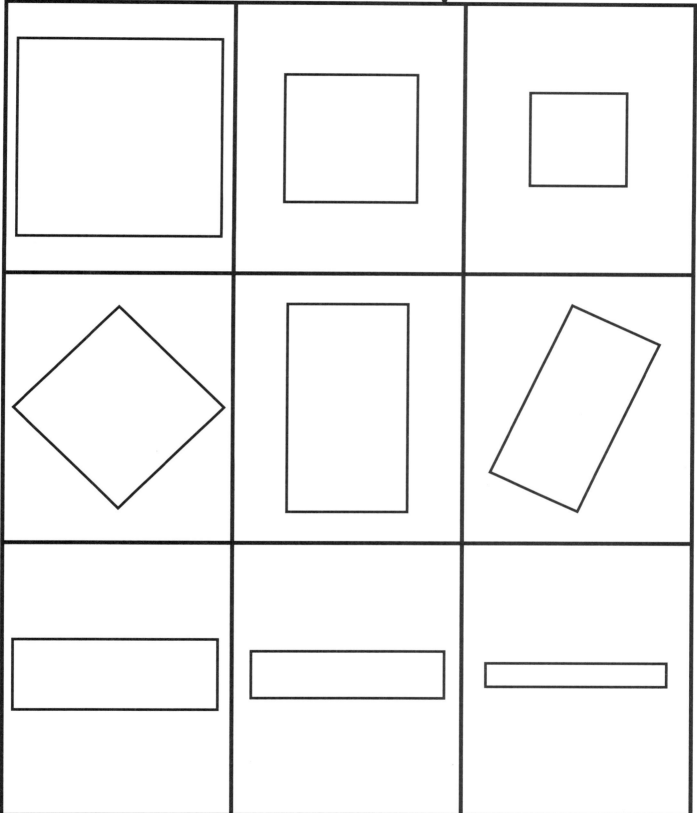

GETTING INTO GEOMETRY © 2010 AIMS Education Foundation

 **Shape Cards**

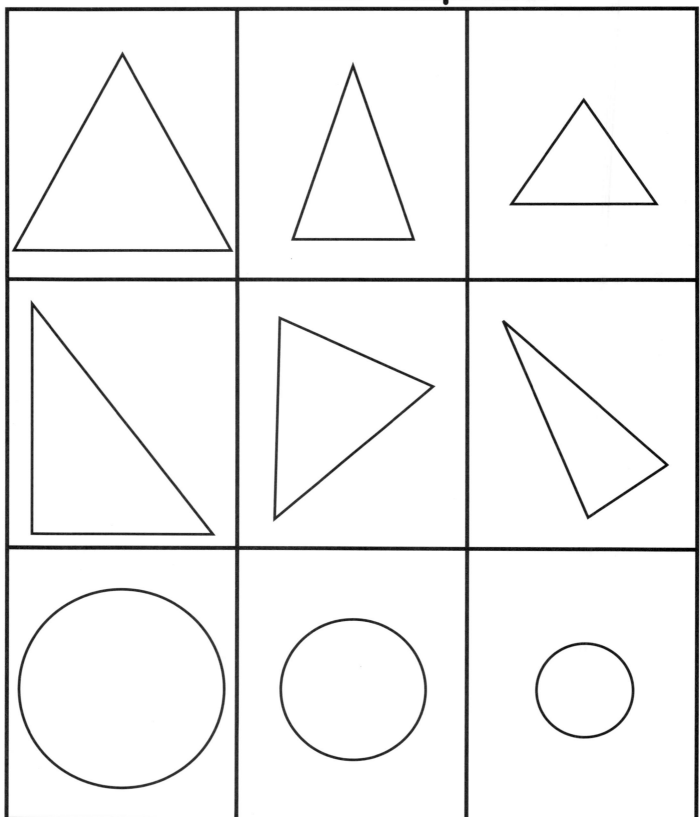

GETTING INTO GEOMETRY © 2010 AIMS Education Foundation

Stepping Out Shapes

Topic
2-D shapes

Key Question
How can stepping around a shape on the playground help you understand more about shapes?

Learning Goals
Students will:
- step around chalk shapes on the playground, and
- review the properties of 2-D shapes including edges and corners.

Guiding Documents
Project 2061 Benchmarks
- *Numbers and shapes can be used to tell about things.*
- *Shapes such as circles, squares, and triangles can be used to describe many things that can be seen.*

*NCTM Standards 2000**
- *Recognize, name, build, draw, compare, and sort two- and three-dimensional shapes*
- *Describe attributes and parts of two- and three-dimensional shapes*
- *Recognize and represent shapes from different perspectives*

Math
Geometry
    2-D shapes
        characteristics
        identification

Integrated Processes
Observing
Comparing and contrasting
Communicating

Materials
Sidewalk chalk (see *Management 1*)
Pictures of geometric shapes (see *Management 3*)
CD player, optional (see *Management 4*)

Background Information
Many young children need kinesthetic activity to help reinforce their understanding of math concepts. In this activity, the focus is on the attributes of two-dimensional geometric shapes. The children will step around the permieters of several shapes drawn on the playground and recognize the numbers of edges and corners they have.

Management
1. Prior to beginning this activity, use chalk to draw a square, triangle, and rectangle on the blacktop or sidewalk. The shapes should be large enough for the entire class to stand around.
2. Locate a large painted circle on the school grounds, if possible. If there isn't one, draw a circle, too.
3. Copy the pictures of the geometric shapes provided to show the class as they prepare to walk around each shape.
4. If desired, use a CD player to add music to the activity. Moving to music may add interest to the shape step activity and support those children with musical strengths.

Procedure
1. Show the included pictures of geometric shapes to the class. Review the shapes with your students allowing them to identify the names and attributes of the square, rectangle, triangle, and circle.
2. Walk with your class to the predetermined playground area and point out the various shapes you have drawn. Question the students about the properties of the various shapes and instruct individual students to stand on the corners and edges of the different shapes.
3. Choose one of the large shapes and instruct the students to stand on the chalk line. Tell them that they will walk on the line around the shape. Remind them to stay well back from the person in front of them to avoid stepping on him/her. Give a signal to begin and have the students step around the shape. Accompany the movement with music, if desired.
4. Take the group to a different shape and review the properties of the shape and directions for moving around the shape.

GETTING INTO GEOMETRY   © 2010 AIMS Education Foundation

5. Repeat until each of the shapes has been used and the students have had adequate practice identifying the properties of each shape.

Connecting Learning
1. What are the names of the shapes that we stepped around on the playground?
2. Were any of the shapes easier to move around than others? Explain.
3. Did stepping around the shapes give you a better understanding of shapes? Explain.
4. How were you able to recognize if you were on an edge or a corner of each shape?
5. How many corners did the triangle have? How many edges?
6. What shapes have four edges? How are they different?
7. How is a circle different from a rectangle and a triangle?
8. How many corners does a circle have?
9. Can you think of any other shapes that we could draw and step around? What are the properties of those shapes? Will they be harder, easier, or the same difficulty to step around? Explain.

* Reprinted with permission from *Principles and Standards for School Mathematics*, 2000 by the National Council of Teachers of Mathematics. All rights reserved.

GETTING INTO GEOMETRY

Square

Rectangle

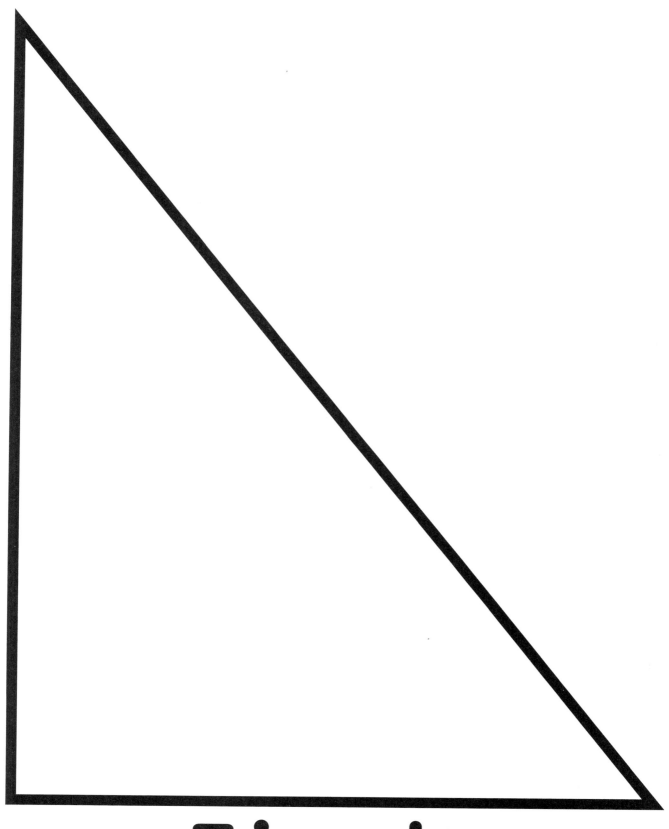

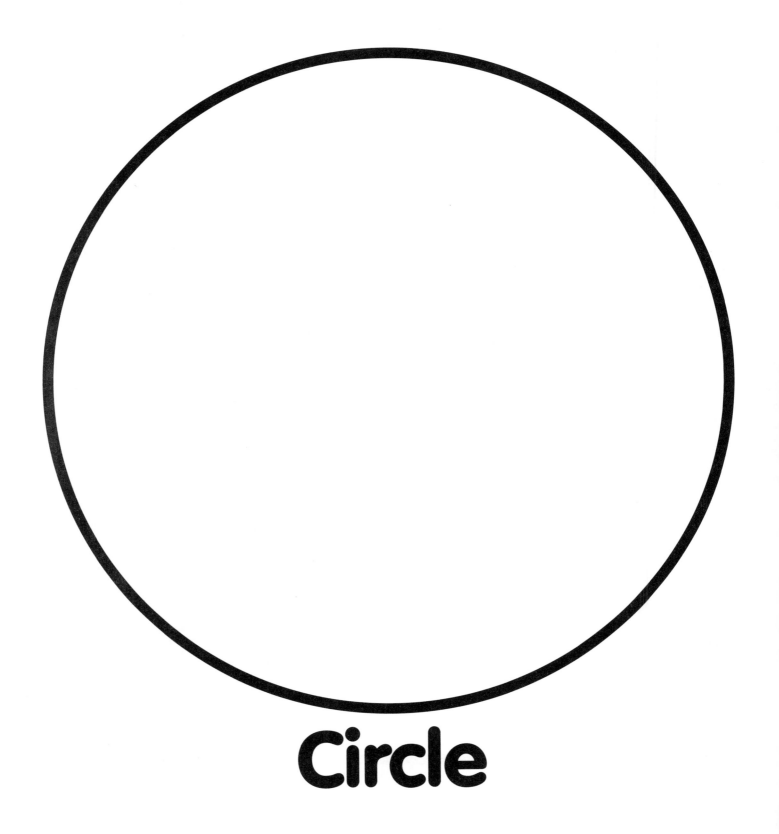

Art in the Air

Topic
2-D shapes

Key Question
How can we use streamers to look at properties of shapes?

Learning Goals
Students will:
- use streamers to make large shapes in the air, and
- review properties of shapes.

Guiding Documents
Project 2061 Benchmarks
- *Numbers and shapes can be used to tell about things.*
- *Shapes such as circles, squares, and triangles can be used to describe many things that can be seen.*

NCTM Standards 2000*
- *Recognize, name, build, draw, compare, and sort two- and three-dimensional shapes*
- *Describe attributes and parts of two- and three-dimensional shapes*
- *Create mental images of geometric shapes using spatial memory and spatial visualization*
- *Recognize and represent shapes from different perspectives*

Math
Geometry
   2-D shapes
      characteristics

Integrated Processes
Observing
Comparing and contrasting
Communicating

Materials
Crepe paper or scarves (see *Management 1*)
Pictures of shapes, optional (see *Management 3*)
CD player, optional

Background Information
    Young children can benefit from gross motor activity when learning a concept. This activity is designed to reinforce the properties of two-dimensional geometric shapes while providing for gross motor development. The activity adds to the kinesthetic support young children need as they learn and reinforce various attributes of shapes.

Management
1. Cut brightly colored crepe paper streamers into three-foot lengths, or gather scarves that are lightweight and long.
2. The use of music is suggested to add interest and encourage movement. Select music with an upbeat tempo to play while the students are making their shapes in the air.
3. If necessary for your students, provide pictures of the shapes to be used.
4. During the activity, have the students change from one hand to another. This exercises the muscles on both sides of the body.
5. As the students are making the shapes, instruct them to make them big so that they go from one side of their body to the other.

Procedure
1. Lead students into an area where they can form a large circle. Instruct them to put their arms out wide and to be sure they don't touch anyone else. Adjust the circle so all students have enough space.
2. Distribute the paper streamers or scarves. Allow the students to experiment with swishing the streamers in front of their bodies. Encourage them to make big shapes, tiny shapes, etc.
3. Inform the class that they will be using the streamers to make the shapes they have been studying. Ask students to name some shapes and describe their attributes (number of edges, corners, etc.).
4. Tell them you will say a shape (and show them a picture of the shape, if necessary) and they will use the paper streamers to make the shape in front of their bodies as large as their arms will let them. Tell them that each time they hear a new shape, they will change their movements to match the properties of the new shape.
5. Play the music (if available), say the name of a shape, and watch as the students begin to swirl the streamers. When all of the students are successful making the shape, change to another shape.

6. After a few minutes of making shapes with one hand/arm, direct the students to change hands and continue to follow the same directions.

Connecting Learning
1. What shapes did you make in the air?
2. Which shape was the easiest to make? Explain.
3. When you made a square, what did you do? [made four lines, four corners, etc.] ...a triangle? ...rectangle? ...etc.?
4. Was it easier to make the shapes with your right hand/arm or your left? Explain.
5. Can you think of another shape that you could make in the air besides the geometric shapes? Explain.

* Reprinted with permission from *Principles and Standards for School Mathematics*, 2000 by the National Council of Teachers of Mathematics. All rights reserved.

Sid the Snake

Topic
2-D shapes

Key Question
What can we learn about shapes by making them with chenille stem "snakes"?

Learning Goals
Students will:
- identify basic two-dimensional shapes,
- bend a chenille stem to create specific shapes, and
- identify characteristics of those shapes.

Guiding Documents
Project 2061 Benchmarks
- *Numbers and shapes can be used to tell about things.*
- *Shapes such as circles, squares, and triangles can be used to describe many things that can be seen.*

NCTM Standards 2000*
- *Recognize, name, build, draw, compare, and sort two- and three-dimensional shapes*
- *Describe attributes and parts of two- and three-dimensional shapes*
- *Recognize and represent shapes from different perspectives*

Math
Geometry
    2-D shapes
        characteristics
        identification

Integrated Processes
Observing
Identifying
Comparing and contrasting
Communicating
Recording

Materials
For each student:
    one chenille stem
    Sid the Snake Makes Shapes book

Background Information
It is important for young children to be able to identify the basic shapes and describe them based on characteristics such as number of edges and corners, whether they are made of straight lines or curved lines, etc. In this activity, students will identify shapes made by a cartoon snake, discuss the characteristics of the shapes, and then make their own versions of the shapes using chenille stem "snakes."

Management
1. Copy the *Sid the Snake Makes Shapes* book pages for each student. Cut the pages in half, put them in order, and staple along the left edge.
2. To reduce the number of copies, you may choose to display the book pages using a projection device rather than giving each student a copy of the book.

Procedure
1. Show the class "Sid the Snake" (a chenille stem). Tell the class that Sid can do tricks. Explain that he can make shapes.
2. Invite a student to name a shape for Sid to make. Make the shape with the chenille stem and place it so that the students can see Sid's new shape. Discuss what it is called, how many edges and corners it has, etc.
3. Tell the class that they are going read a book about Sid and some of the shapes he can make.
4. Give each student a copy of the *Sid the Snake Makes Shapes* book and a chenille stem (or, display the book pages as suggested in *Management 2*).
5. Read the first page of the book together. Ask students to identify the shape on page one. Read the second page and have students make triangles with their snakes. Discuss the number of edges and corners and whether all triangles must look the same.
6. Repeat this process for the remaining pages of the book.
7. On the final page, have students draw some other shapes that could be made with the snakes. Encourage students to describe the characteristics of the shapes as well as name them.

Connecting Learning
1. What shapes did Sid make?
2. Which shape was the easiest to make with your "snake"? Why? ...hardest? Why?
3. What would Sid look like if he were making a triangle? [three corners, three edges] Did all of our triangles look the same? [No.] How did we know they were all triangles if they didn't look the same? [They all had three corners and three edges.]
4. What other shapes could we make with our "snakes"?
5. How does making shapes with "snakes" help us learn about shapes?

* Reprinted with permission from *Principles and Standards for School Mathematics*, 2000 by the National Council of Teachers of Mathematics. All rights reserved.

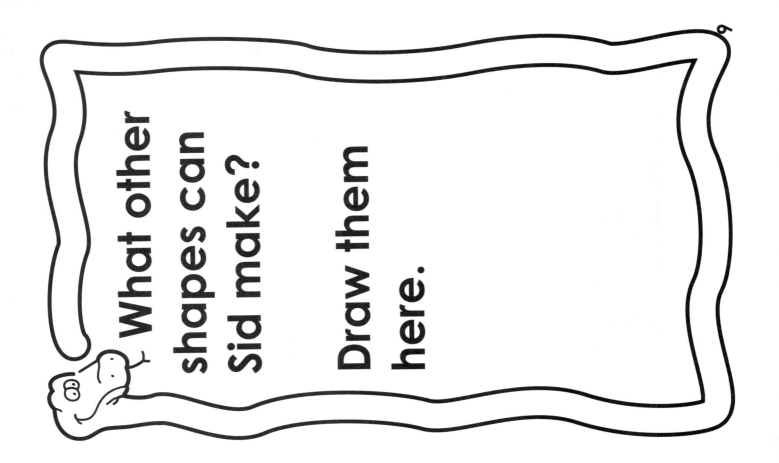

What other shapes can Sid make?

Draw them here.

Sid the Snake Makes Shapes

It is a triangle.

Make a triangle with your snake.

How many edges? _____

How many corners? _____

2

Sid the snake can make this shape.

What shape is it?

1

GETTING INTO GEOMETRY   31   © 2010 AIMS Education Foundation

It is a circle.

Make a circle with your snake.

How many edges? _____

How many corners? _____

4

Sid the snake can make this shape.

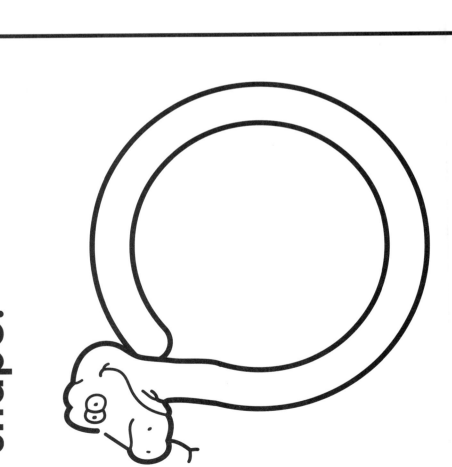

What shape is it? 3

GETTING INTO GEOMETRY © 2010 AIMS Education Foundation

It is a square.

Make a square with your snake.

How many edges? _____

How many corners? _____

Sid the snake can make this shape.

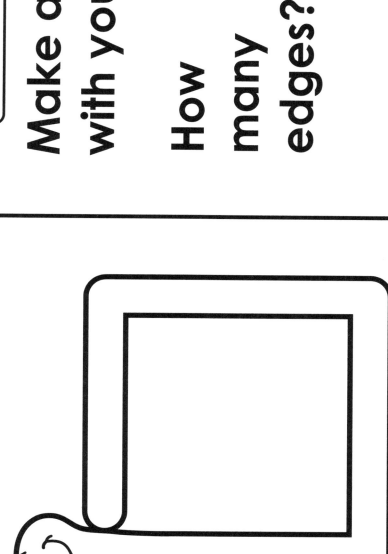

What shape is it?

It is a rectangle.

Make a rectangle with your snake.

How many edges?

How many corners?

Sid the snake can make this shape.

What shape is it?

Sandy Shapes

Topic
2-D shapes

Key Question
How can you identify shapes without seeing them?

Learning Goal
Students will use their sense of touch to identify various squares, rectangles, and triangles.

Guiding Documents
Project 2061 Benchmark
* *Numbers and shapes can be used to tell about things.*

NCTM Standards 2000*
* *Recognize, name, build, draw, compare, and sort two- and three-dimensional shapes*
* *Describe attributes and parts of two- and three-dimensional shapes*
* *Create mental images of geometric shapes using spatial memory and spatial visualization*
* *Recognize and represent shapes from different perspectives*

Math
Geometry
   2-D shapes
      characteristics
      identification

Integrated Processes
Observing
Comparing and contrasting
Communicating

Materials
Disposable bowls
White glue
Sand
Plastic coffee stirrers
Scratch paper, optional (see *Management 4*)

Background Information
    It is important for young children to have multiple experiences identifying two-dimensional shapes. This lesson allows them to use their sense of touch to observe and identify two-dimensional shapes by the number of edges and corners they have.

Management
1. Prior to teaching this lesson, prepare several shape bowls. Cut coffee stirrers into short lengths and use these to make various triangle, square, and rectangle outlines in the bottom of each bowl. Glue the coffee stirrers in place in the bottoms of the bowls. Use various sizes and orientations of the shapes. For example, include a right triangle, scalene triangle, and isosceles triangle, as well as tall thin rectangles, short wide rectangles, etc. Prepare the bowls a day or two in advance so the glue can dry.

2. Label the rim of each bowl with a letter and create a shape key so that the bowls can be self-checking.
3. Cover the dried geometric shapes with a layer of sand. Rice, lentils, or other grains can be substituted for the sand; however, the coffee stirrer shape may not be as pronounced in these items.
4. The recording piece of this activity is optional; if your students' writing and drawing skills are weak, ask them to verbally identify the shapes only.

Procedure
1. Draw various shapes on the board and ask the students to identify them. Ask the class if they think they could recognize these shapes without using their sense of sight.
2. Discuss how they might use their sense of touch to identify the various shapes.
3. Display one of the prepared bowls containing a coffee stirrer shape covered in sand. Explain to the students that they are going to "dig for buried treasures." Tell the class that the buried treasure is actually a secret shape that they will be trying to identify using only their hands.

GETTING INTO GEOMETRY

4. Give each group of students a bowl and piece of scratch paper. Have students label their papers using the letters from the prepared bowls.
5. Ask the students to draw the shape they feel next to the corresponding letter. When groups have identified the shapes in their bowls, have them exchange bowls and identify the new shape. Repeat until all groups have had a chance to identify all shapes.
6. When the rotation of bowls is completed, compare what shape each group found in each bowl. Empty the sand from the bowls and allow students to compare their drawings to the actual shapes.

Connecting Learning
1. Which shapes were easiest to identify? Why?
2. Is it easier to identify the shapes with your eyes or your hands? Explain your thinking.
3. How did you know what shape you were feeling?
4. Could we use the coffee stirrers to make circles in the bowls? Explain your thinking.

* Reprinted with permission from *Principles and Standards for School Mathematics*, 2000 by the National Council of Teachers of Mathematics. All rights reserved.

Rough Enough Shapes

Topic
2-D shapes

Key Question
How can you identify shapes using your sense of touch?

Learning Goal
Students will identify sandpaper shapes using their sense of touch.

Guiding Documents
Project 2061 Benchmark
- *Numbers and shapes can be used to tell about things.*

NCTM Standards 2000*
- *Recognize, name, build, draw, compare, and sort two- and three-dimensional shapes*
- *Describe attributes and parts of two- and three-dimensional shapes*
- *Create mental images of geometric shapes using spatial memory and spatial visualization*
- *Recognize and represent shapes from different perspectives*

Math
Geometry
    2-D shapes
        characteristics
        identification

Integrated Processes
Observing
Comparing and contrasting
Communicating

Materials
For the class:
    sandpaper shapes (see *Management 1*)
    blindfolds, optional (see *Management 2*)

Background Information
    We use information from previous experiences to help us understand our world and to learn about new things. Much of the information is gathered through the use of our five senses. In this activity, students will try to identify four different sandpaper shapes by using only their sense of touch. They will use their prior knowledge of shapes and the sensory input from touch to determine the shapes.

Management
1. Cut a large set of shapes—square, triangle, circle, and rectangle—out of sandpaper. A coarse grade of sandpaper (60-100) is best.
2. Students will need to close their eyes or be blindfolded. To prevent the spread of any eye conditions from one child to another, it is advised that students each have their own felt blindfold. A blindfold pattern has been provided in this activity. Trace the pattern onto a piece of light-colored felt, cut it out, cut a small X in each end, and loop a #19 rubber band through the hole at each end. Use a permanent marker to write the students' names on their blindfolds.

Procedure
1. Place the set of sandpaper shapes on the board tray at the front of the class.
2. Have various students identify the first, second, third, and fourth shape on the tray. Question them about how they know it is a (name of shape). [It had three edges and three corners so I knew it was a triangle. Etc.]
3. Ask the students if they think they could identify the shapes without their sense of sight. Discuss how they would do that. [Use the sense of touch.]
4. Invite four students to come to the front of the class while the others observe. Ask the first student to close his/her eyes or to use a blindfold.
5. Explain that you will be changing the order of the shapes on the tray, and that you would like him/her to identify the shapes along the tray using only his/her sense of touch.
6. Allow the student to slide his/her hand along the tray to the first shape. Have the student remove the shape from the tray and try to identify it.
7. Once the student has declared the identity of the shape, ask questions about why he/she doesn't think it is another shape. The student should respond back using attributes of the shapes, such as, "It is a square because it has four equal edges and four corners."

GETTING INTO GEOMETRY   © 2010 AIMS Education Foundation

8. Repeat the process until all students have had a chance to identify the various shapes.
9. End with a discussion about the strategies used to determine the identity of each shape.

Connecting Learning
1. What sense did you use to identify the shapes the first time? [sight] Was this easy or difficult? Explain.
2. What sense did you use to identify the shapes the second time? [touch] Was this easier or harder than using the sense of sight? Explain.
3. Did you make any errors in identifying the shapes when you were using your sense of touch? Why do you think that happened?
4. Which shape was the easiest for you to identify? Why? Which shape was the most difficult? Why?
5. Can you think of times where you have to identify something by its shape without the use of your sight? [locating something in the dark, searching for an object inside a bag or backpack without looking, etc.]

* Reprinted with permission from *Principles and Standards for School Mathematics,* 2000 by the National Council of Teachers of Mathematics. All rights reserved.

Patterns for Student Blindfolds

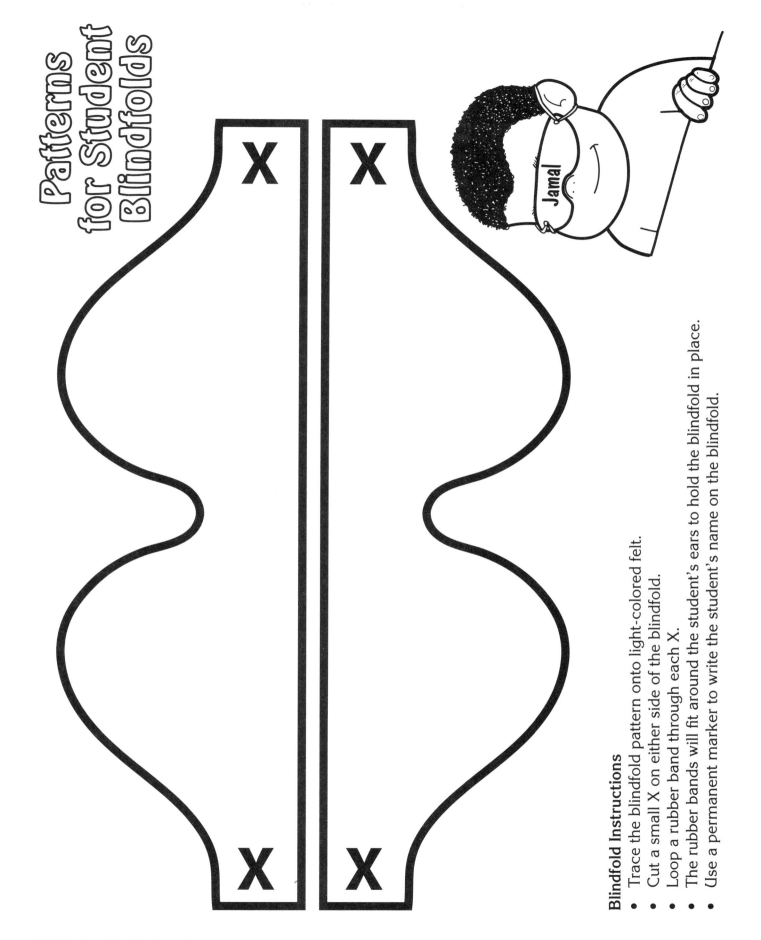

Blindfold Instructions
- Trace the blindfold pattern onto light-colored felt.
- Cut a small X on either side of the blindfold.
- Loop a rubber band through each X.
- The rubber bands will fit around the student's ears to hold the blindfold in place.
- Use a permanent marker to write the student's name on the blindfold.

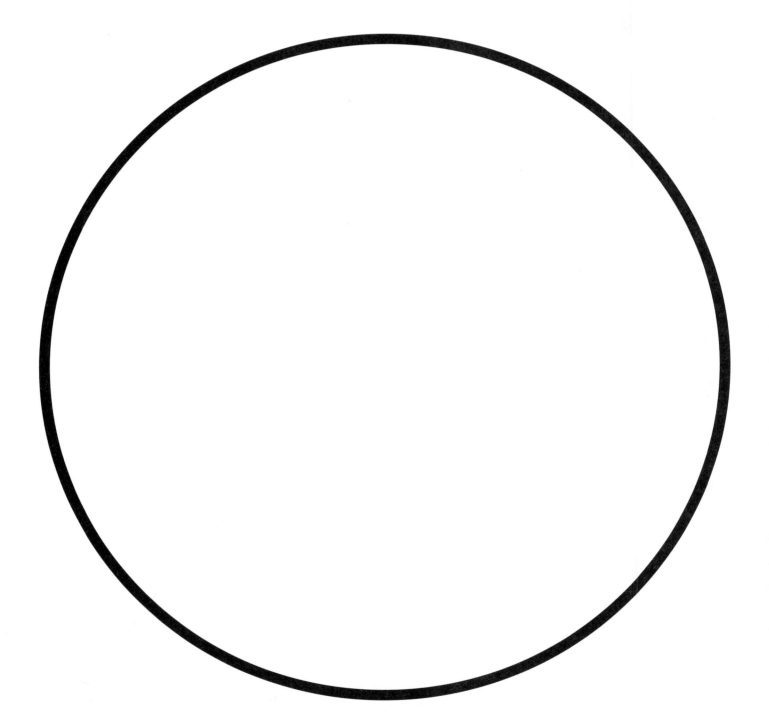

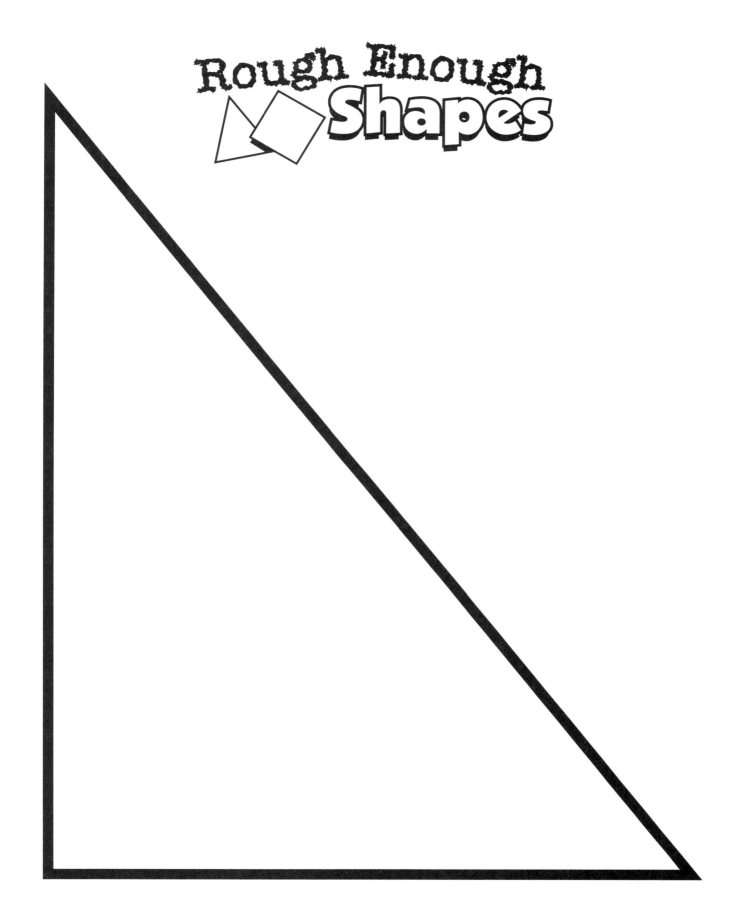

Rough Enough Shapes

GETTING INTO GEOMETRY

Shapes Line Up

Topic
2-D shapes

Key Question
How can we order sets of shapes based on their properties?

Learning Goal
Students will focus on an attribute and order a set of shapes accordingly.

Guiding Documents
Project 2061 Benchmarks
- *Numbers and shapes can be used to tell about things.*
- *Shapes such as circles, squares, and triangles can be used to describe many things that can be seen.*

NCTM Standards 2000*
- *Recognize, name, build, draw, compare, and sort two- and three-dimensional shapes*
- *Describe attributes and parts of two- and three-dimensional shapes*
- *Recognize and represent shapes from different perspectives*

Math
Geometry
   2-D shapes
       characteristics
Ordering

Integrated Processes
Observing
Comparing and contrasting
Communicating

Materials
For the class:
   sets of shapes (see *Management 2*)

For each student:
   shape picture card (see *Management 4*)

Background information
   Ordering is the process of looking at a set of objects—in this case, shapes—focusing on an attribute, and then arranging the set accordingly. For example, a set of children may be ordered from shortest to tallest. Geometric shapes can be ordered according to size, number of corners, number of straight edges, etc.
   Before asking young children to order a set of three or more shapes, it is important that they have multiple opportunities to compare two shapes. This allows the children to recognize attributes by which the objects can be ordered and provides opportunity for them to explore comparative language. For example, the rectangle is *longer* than the square, or the triangle has *fewer* straight edges than the square.
   When children are able to make comparisons, they should work with sets of shapes that vary in size and attributes. Through the ordering experiences provided in this activity, you will be able to observe how your students' observation skills, geometry understanding, logical reasoning abilities, and problem-solving strategies are developing.

Management
1. Cut one large, one small, and one medium triangle out of construction paper for *Part One* of this activity.
2. Gather several sets of 2-D shapes that can be ordered based on specific attributes. These shapes could include pattern blocks, attribute blocks, tangram pieces, foam shapes, etc. There should be enough for one set per group of four students. One set should include shapes that can be ordered by the number of edges. Other sets should include shapes that can be ordered by height, width, number of corners, etc.
3. Divide the class into groups of four for *Part Two* of this activity.
4. Copy a set of the included shape cards onto card stock for *Part Three* of this activity. Laminate for extended use. If there are not enough cards for each student to have one, make additional cards and draw your own shapes on them.

Procedure
Part One
1. Invite two students to the front of the class.
2. Give one student a small triangle and the other a large triangle. Ask the class to make a true statement comparing their triangles. The students might say, " Diego's triangle is taller than Emily's," or "Emily's triangle is shorter than Diego's."

GETTING INTO GEOMETRY   © 2010 AIMS Education Foundation

3. Invite a third student to the front of the class and give him or her a triangle.
4. Ask them to arrange themselves in order from shortest to tallest triangles. Depending on how the triangles are oriented, this order may vary. If there is any question about which is the tallest, etc., have students do a direct comparison.
5. Discuss how they decided which was the shortest, which was the tallest, and which belonged between the two.
6. Select another student from the class and ask him/her to join the group at the front. Encourage the new student to cut a triangle out of construction paper and place it in the correct position based on height.
7. Ask the students about how they decided where the new student's triangle belonged in the order.
8. Discuss real-life occasions when we put things in order based on a specific attribute, such as lining the class up from shortest to tallest on picture day, lining books up on library shelves from tallest to shortest, stacking building blocks largest to smallest, etc.

Part Two
1. Distribute one set of shapes suggested in *Management 2* to each group of students.
2. Ask the group members to observe the shapes and to discuss the attributes by which they could be ordered. [height, number of edges, number of corners, etc.]
3. After discussing the possible attributes by which to order the shapes, ask the groups to chose one attribute and order their shapes accordingly. Have the students use words or pictures to make a record of their chosen order on a blank piece of paper.
4. Have the students trade the sets of shapes several times so that each group of students has an opportunity to order several sets of shapes focusing on different attributes.
5. After each group has been given the opportunity to order several sets of shapes, bring the class together and discuss the different ways that the groups ordered each set of shapes.

Part Three
1. Distribute one shape card to each student.
2. Explain to the students that they will be playing a game that will ask them to order the shape cards the class just received.
3. Tell the class that you will choose a shape and attribute by which the shapes will be ordered. Explain that they should come to the front of the class if they have a picture of the shape mentioned and wait for instructions on how the group should order their shapes. For example, the instruction might be, "Triangles, stand up; line up from largest to smallest."

4. Choose shapes one at a time, and have the students stand up and line up their pictures based on specific attributes. Repeat this process so that each child has an opportunity to participate.
5. After each group has lined up in order, discuss whether they are correctly lined up based on the attribute mentioned.
6. When all students have participated in a line up, discuss how they decided on the correct order.

Connecting Learning
1. What are some ways that sets of shapes can be ordered? [height, width, number of edges, number of corners.]
2. What is one example of ordering in the real world? [books on a shelf, kids on picture day]
3. How did you decide on the order for your group?
4. Which set of shapes was easiest to order? Why?

* Reprinted with permission from *Principles and Standards for School Mathematics*, 2000 by the National Council of Teachers of Mathematics. All rights reserved.

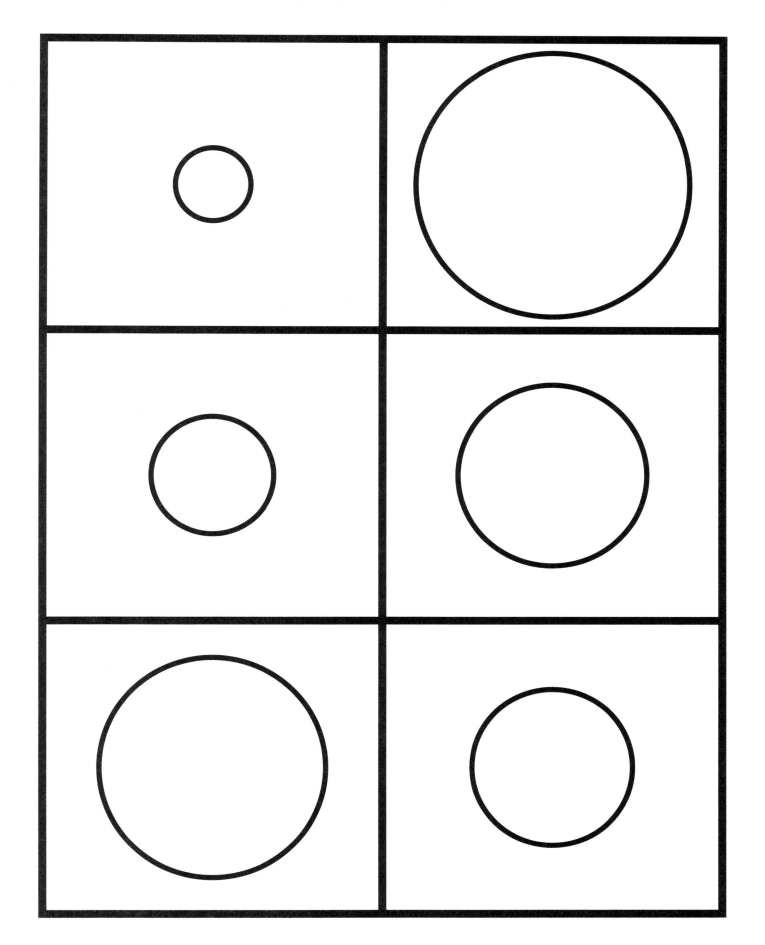

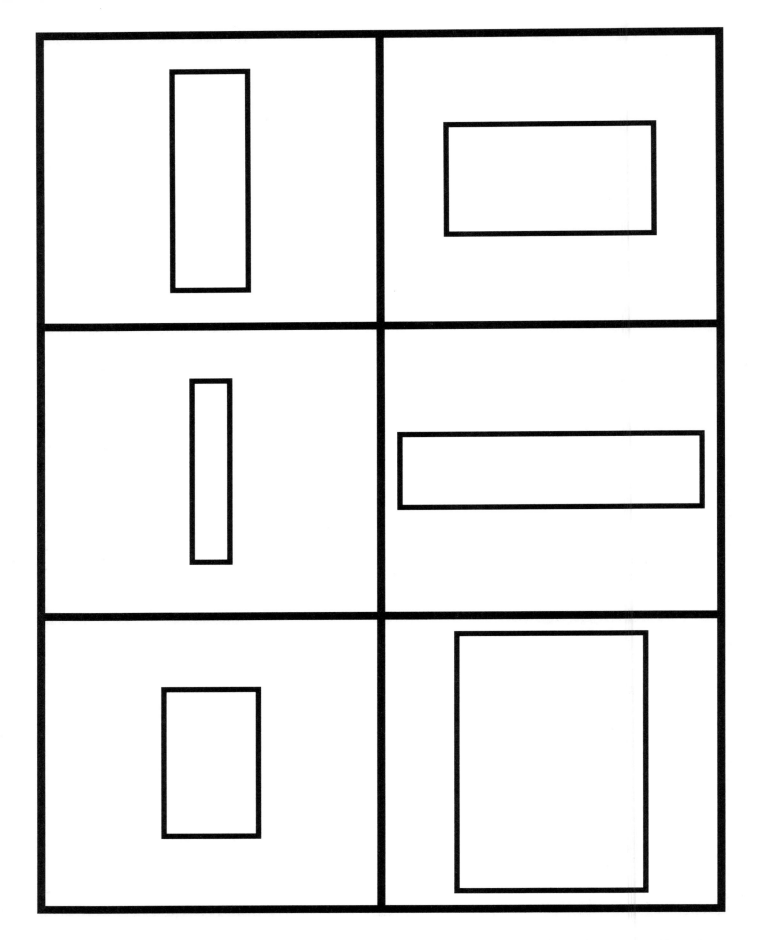

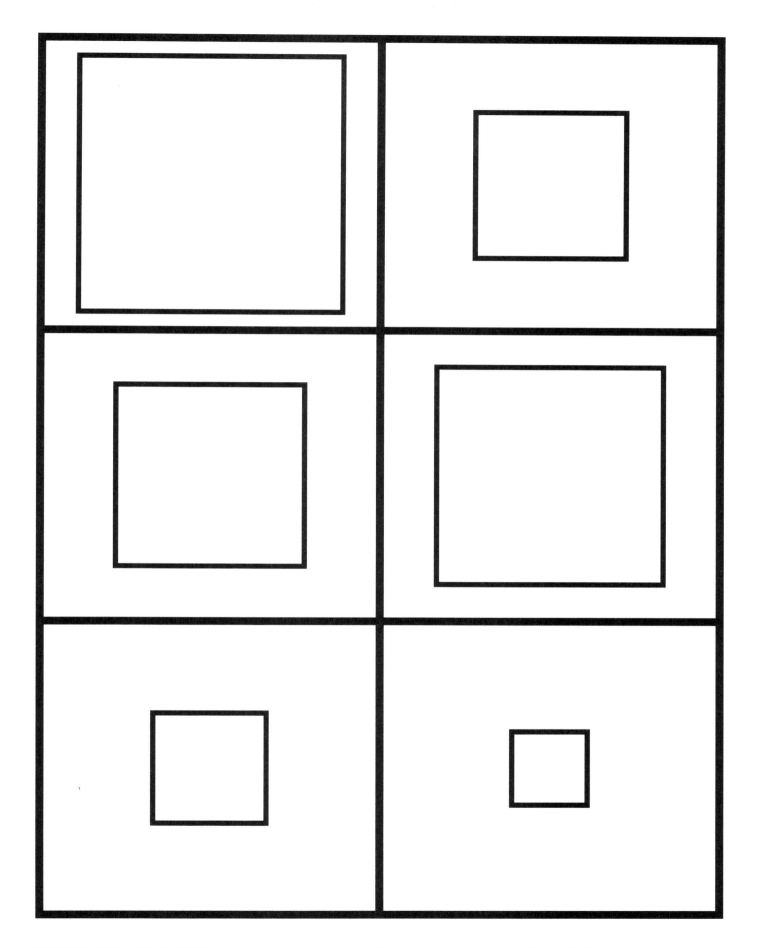

GETTING INTO GEOMETRY 49 © 2010 AIMS Education Foundation

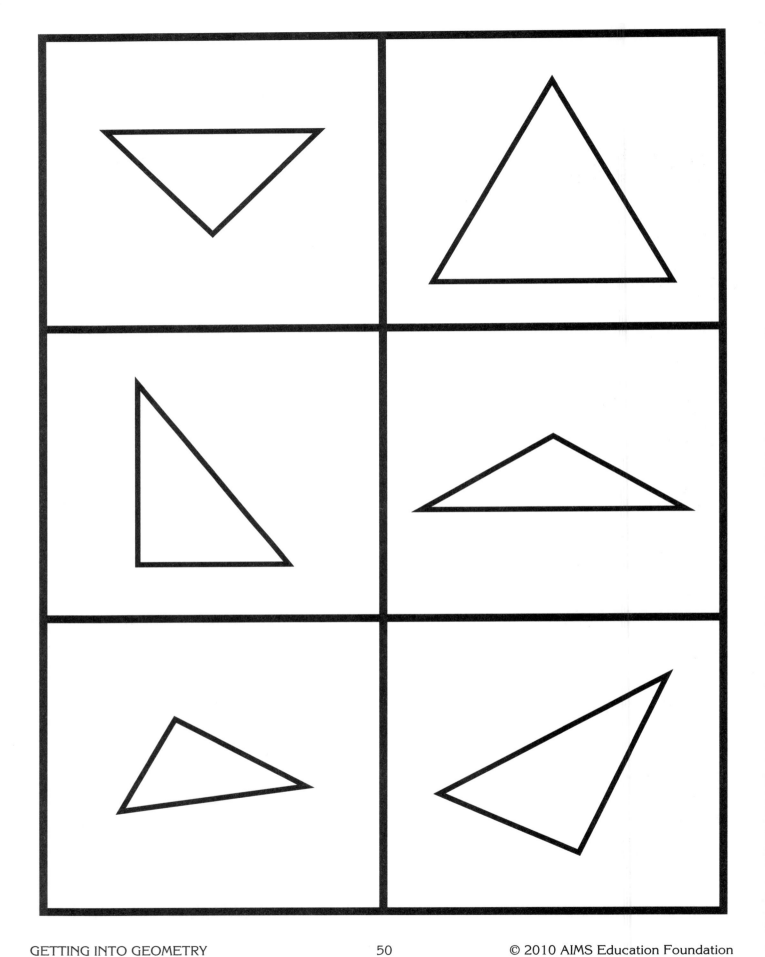

Fabulous Folds

Topic
2-D shapes

Key Question
What shapes can you make by folding a square of fabric?

Learning Goals
Students will:
- recognize and name squares, rectangles, triangles, and circles;
- describe characteristics of these shapes; and
- create these geometric shapes by folding a piece of fabric.

Guiding Documents
Project 2061 Benchmarks
- *Numbers and shapes can be used to tell about things.*
- *Shapes such as circles, squares, and triangles can be used to describe many things that can be seen.*

NCTM Standards 2000*
- *Recognize, name, build, draw, compare, and sort two- and three-dimensional shapes*
- *Describe attributes and parts of two- and three-dimensional shapes*
- *Recognize and represent shapes from different perspectives*

Math
Geometry
  2-D shapes
    characteristics
    modeling

Integrated Processes
Observing
Communicating
Comparing and contrasting

Materials
For each student:
  one 12-inch square of lightweight fabric
    (see *Management 1*)

Background Information
The focus of a study of geometric shapes for early grades should be on concrete experiences in which students observe, touch, and communicate attributes of the shapes. Students should be introduced to different sizes and orientations of geometric shapes in order to understand the concept.

The following is a list of attributes of the geometric shapes covered in this lesson. The list provides true statements that young learners will observe. Students should use their own words for defining and describing these shapes.
- All quadrilaterals, including squares and rectangles, have four edges and four corners.
- All triangles have three edges and three corners.
- All edges on both quadrilaterals and triangles are straight.
- Circles have no corners and no straight edges.
- Turning a shape only changes the orientation, not the attributes of the shape.

Management
1. Use a lightweight fabric such as tulle, organdy, or sheer nylon. Any square scarf that is lightweight can also be used. Material larger than a 12-inch square is difficult to handle.

Procedure
1. Tell students that they will be making shapes by folding pieces of fabric. Model how to fold the fabric for the class. Emphasize the importance of lining up edges and corners.
2. Direct the students to sit on the floor with the fabric laying flat in front of them.
3. Ask students the following geometric and shape questions:
   "What shape is the fabric? "[square]
   "How many edges does it have?" [four]
   "How many corners does it have?" [four]
4. Identify the two top corners and the two bottom corners. Direct students to grab the two top corners with their thumbs and index fingers and fold the top of the fabric down to the two bottom corners to see what shape is formed.
   "Is this new shape a square?" [No.] "Does it have another name?" [rectangle]

GETTING INTO GEOMETRY   51   © 2010 AIMS Education Foundation

"How many edges does the rectangle have?" [four]
"How many corners does it have?" [four]
"How is it like a square?" [four edges, four corners]
"How is it different?" [Two edges are shorter than the other two.]

5. Direct students to fold the rectangle in half from right to left.
"Have you seen this shape before? What is it called?" [square]
"If you now fold from top to bottom again, what shape is formed?" [rectangle]
"How many times do you think you can fold from top to bottom and right to left?"
"How small can you fold your fabric?"

6. Have students unfold the fabric and lay it flat. Have them grab one top corner and fold it across to the opposite bottom corner while reviewing the terms *across* and *opposite*.
"What new shape is formed?" [triangle]
"How many edges does it have?" [three]
"How many corners does it have?" [three]
"How does it compare to a square?" [It is half a square.]

7. Have students fold their triangle shape in half again, folding one corner to another, and ask what shape they got. [triangle]
"Since each fold produces a smaller triangle, how many times can you fold it to make the smallest triangle possible?"

8. Have students unfold the fabric and lay it flat. Challenge them to make a circle from it. After students struggle with the task, discuss why they can't fold the fabric into a circle. [The circle has no straight edges.]

Connecting Learning
1. How many edges are on a square? ...a triangle? ...a rectangle?
2. How many corners are on a circle? ...a square? ...a rectangle? ...a triangle?
3. Why is the circle different from the other three shapes? [It has no straight edges or corners.]
4. What shape could we not make with our fabric? [circle] Do you think we could make one? Explain.

Extensions
1. Allow students the opportunity to fold the scarves with a partner giving the directions. This can be done at a station.
2. Encourage students to fold their scarves into squares, triangles, or rectangles on top of construction paper and trace each shape to record it. If they are able, have students label their shapes. (Squares of paper can be used instead of fabric for this extension.)

Curriculum Correlation
Literature
Campbell, Kathy Kuhtz. *Let's Draw a Bear With Squares.* Powerkids Press. New York. 2004.

Maccarone, Grace. *Three Pigs, One Wolf, and Seven Magic Shapes.* Scholastic, Inc. New York. 1997.

Nelson, Robin. *Circle.* Learner Publishing Group, Inc. Minneapolis, MN. 2004.

Nelson, Robin. *Rectangle.* Learner Publishing Group, Inc. Minneapolis, MN. 2004.

Nelson, Robin. *Square.* Learner Publishing Group, Inc. Minneapolis, MN. 2004.

Nelson, Robin. *Triangle.* Learner Publishing Group, Inc. Minneapolis, MN. 2004.

Van Voorst, Jennifer. *Making Shapes.* Capstone Press. Bloomington, MN. 2003.

Music, Language Arts
Change the words of a familiar tune or poem to incorporate the geometric shapes. For example, use the Hokey Pokey tune, but change the words: "You put your yellow circle in, you put your yellow circle out, you put your yellow circle in, and you shake it all about. You do the hokey pokey and you turn yourself around. That's what it's all about." Verse 2: You put your red triangle in, etc. You can have a variety of colors for each shape, incorporating knowledge of colors into the activity.

Use a poem such as *Brown Bear, Brown Bear* and change it to: "Emily, Emily, what do you see?" "I see a green circle in front of me." (One student places the colored shape down, and asks the question. The second student answers.) Students switch roles and repeat the activity using a new shape of a different color.

* Reprinted with permission from *Principles and Standards for School Mathematics*, 2000 by the National Council of Teachers of Mathematics. All rights reserved.

Find the Fit

Topic
2-D shapes

Key Question
What are the attributes of the shapes you have, and where do they belong in the sheet of wrapping paper?

Learning Goals
Students will:
- identify shapes and their attributes,
- locate the positions in a sheet of wrapping paper from which the shapes were taken, and
- identify the shape within their sets that does not belong.

Guiding Documents
Project 2061 Benchmarks
- *Numbers and shapes can be used to tell about things.*
- *Shapes such as circles, squares, and triangles can be used to describe many things that can be seen.*

NCTM Standards 2000*
- *Recognize, name, build, draw, compare, and sort two- and three-dimensional shapes*
- *Describe attributes and parts of two- and three-dimensional shapes*
- *Recognize and represent shapes from different perspectives*
- *Build new mathematical knowledge through problem solving*
- *Apply and adapt a variety of appropriate strategies to solve problems*

Math
Geometry
    2-D shapes
        characteristics
    spatial sense
Problem solving

Integrated Processes
Observing
Identifying
Comparing and contrasting

Materials
Printed wrapping paper (see *Management 1*)
Envelopes

Background Information
Spatial sense is an important skill in geometry. Students need to be able to recognize shapes in various orientations. For young learners, experience with physical models of shapes can help to develop spatial skills. Not only can they see the shapes, they can manipulate them and experience the results. In this activity, students are challenged to use their spatial skills to determine where cutout shapes fit into a piece of wrapping paper. This involves not only matching the shape to the hole, but also determining the correct orientation so that the pattern of the wrapping paper matches as well.

Management
1. Select a wrapping paper that has a somewhat busy pattern, not one that is uniform or geometric. Each student group will need one sheet of wrapping paper large enough to have several shapes cut out of it. If using rolled paper, provide each group with approximately 18" x 24".
2. If possible, laminate the wrapping paper before you cut the shapes out. This provides durability and allows for extended use.
3. Copy the page of shapes provided and use them as templates to trace around and cut out. Select the shapes that are appropriate for your students.
4. To prepare each section of wrapping paper, cut out multiple copies of each shape in different areas of the paper. Be sure that your cuts are uniform, precise, and clean so that it is not obvious which shape came from which hole. (It may be helpful to use a craft knife rather than scissors to make the cuts.) Write a number or letter in one corner of the wrapping paper. Put all the shapes that were cut from that paper into an envelope with a corresponding number or letter. Include one extra shape in each envelope cut from a piece of wrapping paper that will not be given to any group.

Procedure
1. Have students get into groups of three or four. Distribute an envelope of shapes to each group.
2. Instruct students to empty their envelopes and identify the shapes. Have them share attributes of the shapes. [Triangles have three edges/corners; squares and rectangles have four edges/corners; all of a square's edges are the same length; etc.]

GETTING INTO GEOMETRY   © 2010 AIMS Education Foundation

3. If desired, have students sort the shapes by different properties.
4. Distribute to each group the sheet of wrapping paper that corresponds to their shapes.
5. Tell students that their challenge now is to discover where their shapes came from on the page of wrapping paper. Explain that it is not enough to match the shape to a hole, they must also match the pattern of the wrapping paper. Also inform them that they are looking for an imposter—one shape that does not belong on their paper.
6. Allow time for students to find the locations for their shapes. Have them identify the shape that does not belong. If desired, have all groups try to fit their imposter pieces in the sheet of wrapping paper from which they were cut.

Connecting Learning
1. What shapes did you have in your envelope?
2. What are some attributes of those shapes?
3. How is a triangle like a square? How is it different?
4. How is a square like a rectangle? How is it different?
5. What did you have to do to find where each shape belonged in your piece of wrapping paper? [find where both shape and pattern were the same]
6. What things made this difficult?
7. Which shape does not belong? How do you know?

* Reprinted with permission from *Principles and Standards for School Mathematics*, 2000 by the National Council of Teachers of Mathematics. All rights reserved.

Find the Fit

This page can be copied onto card stock and the shapes cut out to use as templates. Trace multiple copies of each of the desired shapes on every group's piece of wrapping paper.

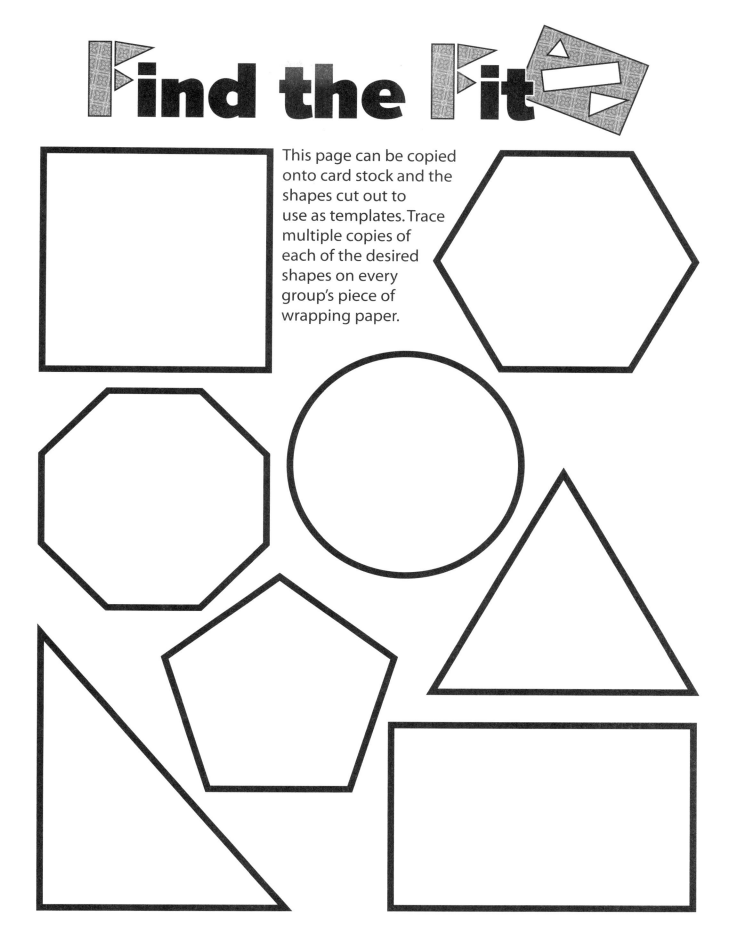

GETTING INTO GEOMETRY © 2010 AIMS Education Foundation

Shapes on the Shelf

Topic
2-D shapes

Key Question
How can we use a storyboard to act out stories using geometric shapes?

Learning Goals
Students will:
- use their own words to describe experiences in a mathematical setting, and
- use manipulatives to represent different scenarios.

Guiding Documents
Project 2061 Benchmarks
- *Numbers and shapes can be used to tell about things.*
- *Shapes such as circles, squares, and triangles can be used to describe many things that can be seen.*

NCTM Standards 2000*
- *Recognize, name, build, draw, compare, and sort two- and three-dimensional shapes*
- *Describe attributes and parts of two- and three-dimensional shapes*
- *Recognize and represent shapes from different perspectives*

Math
Geometry
 2-D shapes
Problem solving

Integrated Processes
Observing
Comparing and contrasting
Relating
Communicating

Materials
Storyboard (see *Management 1*)
Pattern blocks (see *Management 3*)

Background Information
By moving geometric objects around while they are solving a problem, young students develop visual images of the shapes and can more easily see the similarities and differences among them. By taking an active role in the problem and in finding a solution, students are more likely to remember the attributes of the shapes and the relationship between the shapes and be able to solve similar problems when they encounter them again.

Management
1. The storyboard can be colored and laminated for extended use. It can also be displayed using a projection device.
2. While the focus of this activity is on attributes of 2-D shapes, it is also an excellent opportunity to incorporate positional words as students place shapes on the storyboard and describe their locations.
3. This activity uses five different pattern blocks: square, triangle, hexagon, trapezoid, and the blue rhombus. If you do not want to introduce the vocabulary for the hexagon, trapezoid, and/or rhombus shapes, they can be identified by color instead. The tan rhombus shape will not be used.

Procedure
1. Give each student a storyboard and a selection of pattern blocks (see *Management 3*).
2. Have students identify each of the pattern blocks and describe some of its characteristics. [The green shape is a triangle. It has three edges and three corners. The red shape has four edges and four corners. It is called a trapezoid. Etc.]
3. Read one of the scenarios aloud to your students. Then read it again while they act it out with the pattern blocks on their boards. Some of the students may need to listen to the scenario several times while they solve the problem.
4. Tell the students to clear their boards after they finish each scenario.
5. Repeat the process with as many problems as desired.
6. Invite students to make up their own scenarios to share with a partner or the whole class.

Possible Scenarios
- Place one of each shape on the top shelf. How many shapes did you put on the shelf? [5] Which shape in your picture has the most edges? [hexagon (yellow shape)] Which shape in your picture has the fewest corners? [triangle] Describe your picture.
- Put one triangle on the top shelf and one square on the bottom shelf. Add one more shape of your choosing anywhere in the picture. Which of the shapes has the fewest corners? Which of the shapes has the most edges? Describe your picture.

- Put three triangles in your picture. Describe their locations. How many edges are there in all? [9] How many corners are there in all? [9]
- Put the hexagon (the yellow pattern block), on the middle shelf. Put a square on the bottom shelf. Put a triangle on the top shelf above the hexagon. Put another triangle, beside the hexagon. How many corners are there in your picture? [16] How many edges are in your picture? [16] Where is the shape with six edges? [on the middle shelf]
- There are nine edges in the picture. Which pattern blocks could be in the picture? [three triangles; one hexagon and one triangle] How many different ways are there to show nine edges in the picture? [two]
- Place one triangle on the windowsill. Place one trapezoid (red pattern block) in the clothesbasket. Place one square beside the fish bowl. Describe your picture. How many edges and corners are there?
- Place one triangle on the books. Place one hexagon beside the books. Place one rhombus (blue pattern block) on the chair. Tell about the number of edges in your picture. Tell about the number of corners in your picture.
- Place one of each shape with four edges on your picture. Describe their locations. How many shapes do you have on your picture? [3] How many edges are there in the picture? [12] How many corners? [12]

Connecting Learning
1. Which pattern blocks have four edges? [square, blue rombus, red trapezoid]
2. How did you know how many corners you had in your picture?
3. How did you decide how many edges were in the picture?
4. What did you notice about the number of edges and the number of corners in a picture? [They were always the same.]
5. How were the shapes you used alike? How were they different?
6. How many more edges does the yellow hexagon have than the triangle? [3]

* Reprinted with permission from *Principles and Standards for School Mathematics*, 2000 by the National Council of Teachers of Mathematics. All rights reserved.

Goin' On A Shape Hunt

Topic
2-D shapes

Key Question
What shapes can we find in the real world?

Learning Goal
Students will identify 2-D shapes in the environment.

Guiding Documents
Project 2061 Benchmarks
- *Circles, squares, triangles, and other shapes can be found in things in nature and in things that people build.*
- *Numbers and shapes can be used to tell about things.*
- *Shapes such as circles, squares, and triangles can be used to describe many things that can be seen.*

NCTM Standards 2000*
- *Recognize, name, build, draw, compare, and sort two- and three-dimensional shapes*
- *Describe attributes and parts of two- and three-dimensional shapes*
- *Recognize and represent shapes from different perspectives*
- *Recognize geometric shapes and structures in the environment and specify their location*

Math
Geometry
    2-D shapes
        characteristics
        identification

Integrated Processes
Observing
Classifying
Collecting and recording data
Comparing and contrasting
Interpreting data

Materials
Toilet paper tubes
Glue
Cotton balls
Yarn
Hole punch
Shape book
Student booklet
A variety of shaped containers
A shape picture book

Background Information
Shapes are all around us—from our breakfast (round pancakes, square cereal), to our houses (rectangular door, square window), to the street signs around town (triangular yield sign, octagonal stop sign). After reading a book about shapes and reviewing various shapes, students will learn to spot them in the environment. The search for shapes will help them tie together skills such as recognizing shapes and using words and pictures to describe their locations. When children look at shapes in a book, they relate the words to geometric figures. When they find these same shapes around them, they begin to see how math connects to their world, which is just as important as thinking critically, sorting, and naming—all skills they'll use to describe what and where the shape is. By using a playful scenario to identify shapes, the chances that the shape hunting skills will continue long after the game ends are increased.

Management
1. Find an object in the classroom that represents each of the four shapes covered in this lesson—circle, triangle, square, and rectangle.
2. It is suggested that the activity begin with the reading of a shape book. See *Curriculum Correlation* for suggestions.
3. Prior to teaching the lesson, send a note home with students asking families to send in toilet paper tubes. Other cardboard tubes such as from paper towels or waxed paper, can be used; they will need to be cut to the length of toilet paper tubes.
4. Make a model set of binoculars using two toilet paper tubes, a cotton ball, and piece of yarn. Glue the cotton ball between the toilet paper tubes and let it dry. Punch holes in the tubes and tie the yarn through the holes.

Procedure
1. Read one of the shape books suggested in the *Curriculum Correlation* section of this activity or your favorite shape book. As you go through the book, point to and talk about the illustrations.
2. Discuss the fact that shapes are found not only on the pages of these books, but all around us.

GETTING INTO GEOMETRY   © 2010 AIMS Education Foundation

Encourage the students to look around the room. Ask them, "What shape is the clock? ...the table? ...the mirror?" Point out that the book you just read is itself a shape! Discuss shapes that they find outside and in nature. For example flowers, trees, the moon, doors, windows, tires, etc.
3. Explain to the students that they will be going on a shape hunt. Display your paper tube binoculars and tell the students that they will make a set to help them spot the shapes.
4. Distribute the materials and assist students in assembling binoculars.
5. Prepare for your hunt by reciting the *Shape Hunt Chant*. Repeat the chant for each shape, holding up the example you've prepared for reference. Invite the students to take part by pausing after the line "Oh, look! A ___ in front of me." Then hold up a shape and let a child call out its name.
6. Take the students outdoors. Tell them that they are now ready to become shape hunters! As you sing the shape hunt chant again, explore the world around you, letting the students take the lead. When you reach the line that says, "Oh, look! A circle in front of me," ask a student to find and point to a circle in the environment. Do the same with each new shape that you have introduced.
7. Bring the children back indoors. Distribute the shape books and ask the children to draw and/or write about what real-world shapes they saw and where they were located.
8. End with a discussion about their findings, whether they found more of one shape than another, and what shapes they might find at home.

Connecting Learning
1. What shapes did you find? Where were they located?
2. What shape did you find more of? Why do you think that is true?
3. What other areas might we find a lot of shapes? Why?
4. Show me a square. ...circle. ...triangle. ...square. ...rectangle.

Curriculum Correlation
Jonas, Ann. *Round Trip*. Greenwillow Books. New York. 1990.

Thong, Roseanne. *Round Is a Mooncake: A Book of Shapes*. Chronicle Books. San Francisco. 2000.

Whitford, Ann Paul. *Eight Hands Round: A Patchwork Alphabet*. HarperCollins. New York. 1996.

* Reprinted with permission from *Principles and Standards for School Mathematics, 2000* by the National Council of Teachers of Mathematics. All rights reserved.

Shape Hunt Chant

Goin' on a shape hunt (repeat)
Leavin' right away (repeat)
Got some binoculars (repeat)
'Round my neck (repeat)
Oh, oh! (repeat)
What do I see? (repeat)
Oh, look! A circle in front of me! (repeat)
It has no edges. (repeat)
It has no corners. (repeat)
Better just draw it. (repeat) *Make motions of drawing*

Repeat with additional shapes:

Square
Oh, look! A square in front of me! (repeat)
It has four corners. (repeat)
It has four equal edges. (repeat)
Better just draw it. (repeat) *Make motions of drawing*

Triangle
Oh, look! A triangle in front of me! (repeat)
It has three edges. (repeat)
It has three corners. (repeat)
Better just draw it. (repeat)
Make motions of drawing

Rectangle
Oh, look! A rectangle in front of me! (repeat)
It has four edges. (repeat)
It has four corners. (repeat)
Better just draw it. (repeat)
Make motions of drawing

GOIN' ON A SHAPE HUNT

GOIN' ON A SHAPE HUNT

What do I see?
I see this square in front of me.

It has four equal edges.
It has four corners.

What do I see?
I see this circle in front of me.

It has no edges.
It has no corners.

What do I see?
I see this triangle in front of me.

It has three edges.
It has three corners.

What do I see?
I see this rectangle in front of me.

It has four edges.
It has four corners.

Shapes on the Bus

Topic
2-D shapes

Key Question
What parts of a school bus are shaped like squares, rectangles, circles, and triangles?

Learning Goals
Students will:
- discover and chart all the circles, squares, rectangles, and triangles they can find on their school bus; and
- be introduced to the safety issues that relate to the various parts they have identified.

Guiding Documents
Project 2061 Benchmarks
- *Circles, squares, triangles, and other shapes can be found in things in nature and in things that people build.*
- *Numbers and shapes can be used to tell about things.*
- *Shapes such as circles, squares, and triangles can be used to describe many things that can be seen.*

NRC Standard
- *Safety and security are basic needs of humans. Safety involves freedom from danger, risk, or injury. Student understandings included following safety rules for home and school, preventing abuse and neglect, avoiding injury, etc.*

NCTM Standards 2000*
- *Recognize, name, build, draw, compare, and sort two- and three-dimensional shapes*
- *Describe attributes and parts of two- and three-dimensional shapes*
- *Recognize and represent shapes from different perspectives*
- *Recognize geometric shapes and structures in the environment and specify their location*

Math
Geometry
   2-D shapes
      identification

Integrated Processes
Observing
Classifying
Collecting and recording data
Communicating

Materials
For the class:
   Shapes on the Bus chart (see *Management 4*)
   sentence strips
   loaf of bread (square)
   cream cheese
   yellow food coloring
   small square cheese crackers
   small round crackers
   small round red candies
   plastic knives

For each student:
   Shapes on the Bus recording book
   scissors
   crayons

For each pair of students:
   metal ring with shapes (see *Management 3*)

Background Information
    Teaching and using mathematics in a real-world situation makes the study of mathematics exciting and inviting to primary students. Using a familiar object ties in previous experiences and furthers their understanding. The actual school bus in this activity provides the context for discovering geometry in the everyday world and for collecting, recording, and organizing data. A word bank is provided to give the students a way of accessing correct terminology for the bus parts. The bus driver will use this list as s/he takes the students on a tour of the bus, identifying the parts of the bus and making connections to the safety issues involved with each.

Management
1. Follow the appropriate procedure for obtaining a school bus for approximately an hour. Request that the bus park in a safe location where traffic will not be a problem and where the children will have space to move around it safely.

GETTING INTO GEOMETRY   © 2010 AIMS Education Foundation

2. Ask the bus driver to give a tour of the bus, identifying the parts listed on *Parts of My School Bus*, and specifically making connections to the safety issues involved with each. Give the bus driver plenty of advance time to prepare for the tour. Depending on your bus driver, you may need to write out more fully the safety connections you would like him/her to make.
3. Copy the page of shapes (circle, square, rectangle and triangle) onto various colors of card stock. Single punch each and put on a metal ring. Make one set for each pair of students.
4. Make the *Shapes on the Bus* chart by enlarging the provided page to fill a piece of chart paper.

5. Assemble the *Shapes on the Bus* recording books as follows:
   - Use three sheets of paper stacked one on top of the other. The bottom sheet should be the *Parts of My School Bus* page. Stagger the sheets so the top one is about one inch higher than the second, and the second sheet so that it is about one inch higher than the bottom sheet.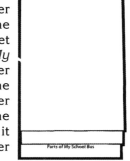
   - Fold as illustrated so that all layers have about a one inch difference from the preceding layer. Leave enough room for the title to fit on the top layer. Staple at the fold.

6. For *Part Two,* use yellow food coloring to tint a container of spreadable cream cheese.

Procedure
Part One: Finding Shapes on a Bus
1. Read a book about school buses prior to the arrival of your school bus (see *Curriculum Correlation* for suggestions).
2. Direct the students to find a partner and distribute the shape rings, one to each set of partners.
3. Review the shapes on the ring and let students take turns identifying shapes in the room by holding up the corresponding shape on the ring when you say the name of the object. (For example, when you say, "clock," students should hold up the circle shape.)
4. Tell the students that they are about to go on a tour of the school bus. Ask, "Do you think we will discover parts of our school bus that have these shapes?"
5. Introduce the *Shapes on the Bus* recording books. Have each student glue the title on the top layer. Have students draw a circle on the flap of the second layer, a square on the flap of the third, a rectangle on the fourth, and a triangle on the fifth. The sixth flap will already be labeled *Parts of My School Bus*.

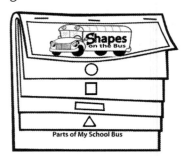

6. Walk the students through the word bank found on the last page of the recording book, identifying the pictures and previewing the words. Direct the students to use the list as a reference for needed words as they explore the bus looking for shapes.
7. Model the use of the recording book with at least two examples, drawing a picture of the bus part on the appropriate shape page and writing the word identifying it underneath.
8. Go on the bus tour. Let the bus driver introduce the class to each part of the bus, making sure that the word bank given in *Parts of My School Bus* is covered, along with a clear connection to bus safety.
9. Direct students to hunt for shapes on the bus with their partners. Each pair can take turns holding the shapes on a ring while the other records their findings.

10. Compile the students' findings on the class chart, *Shapes on the Bus*. As a student shares a part of the bus s/he discovered and its shape, write it on a section of sentence strip and let the student illustrate it and put it in the appropriate column on the chart paper bus.
11. Discuss the results.
12. Review the chart and discuss any safety tips that relate to the parts of the bus.

Part Two: Making an Edible Shape Bus
1. Distribute the page of directions for the edible shape bus.
2. Read through the instructions while making a sample bus in front of the class.
3. Distribute the materials and allow time for students to make their own buses.
4. Have them identify the shapes that make up their edible buses and compare those to the shapes they found in the real bus.
5. Allow students to eat their school buses.

Connecting Learning
Part One
1. What did you find on the bus that was in the shape of a circle? …a square? …a rectangle? …a triangle?
2. How many _____ (shape) did we find?
3. Which shape did we find most of? …least?
4. Are there more (or less) _____ or more (or less) _____?
5. How many more (or less) _____ are there than _____?
6. Are any columns on our chart the same? What does that tell us?
7. How do some of the parts of the bus help keep us safe while we travel?
8. What are some things you need to do when you ride on the bus to keep yourself and others safe?

Part Two
1. What shapes are there on your edible school bus?
2. Which shape do you have the most of on your edible bus?
3. How do these shapes compare to the shapes on the real school bus?
4. Which shape on your edible school bus was your favorite to eat? Why?

Extensions
1. Write a class book using the cloze sentence, "On a bus, a _____ can be a _____."
2. Use a bus to count the number of wheels, the number of seats, the number of windows, the number of doors, the number of steps, etc.

Curriculum Correlation
Literature
Crews, Donald. *School Bus*. HarperTrophy. New York. 1993.

Parker, Marjorie Blain. *Hello, School Bus!* Scholastic, Inc. New York. 2004.

Roth, Carol. *The Little School Bus*. North-South Books. New York. 2002.

Singer, Marilyn. *I'm Your Bus*. Scholastic Press. New York. 2009.

Music
Sing *The Wheels on the Bus*, using the singable books by Raffi or Maryann Kovalski.

Art
Have students draw and share bus safety posters.

Physical Education
Build a bus with blocks and chairs and enact *The Wheels on the Bus*.

Social Studies
Make a book of *Our Community Helpers*, including the bus driver and others you might want to bring in as guests to talk about their jobs (the mail carrier for the school, your school nurse, your librarian, the area firemen and police patrol, etc.).

* Reprinted with permission from *Principles and Standards for School Mathematics, 2000* by the National Council of Teachers of Mathematics. All rights reserved.

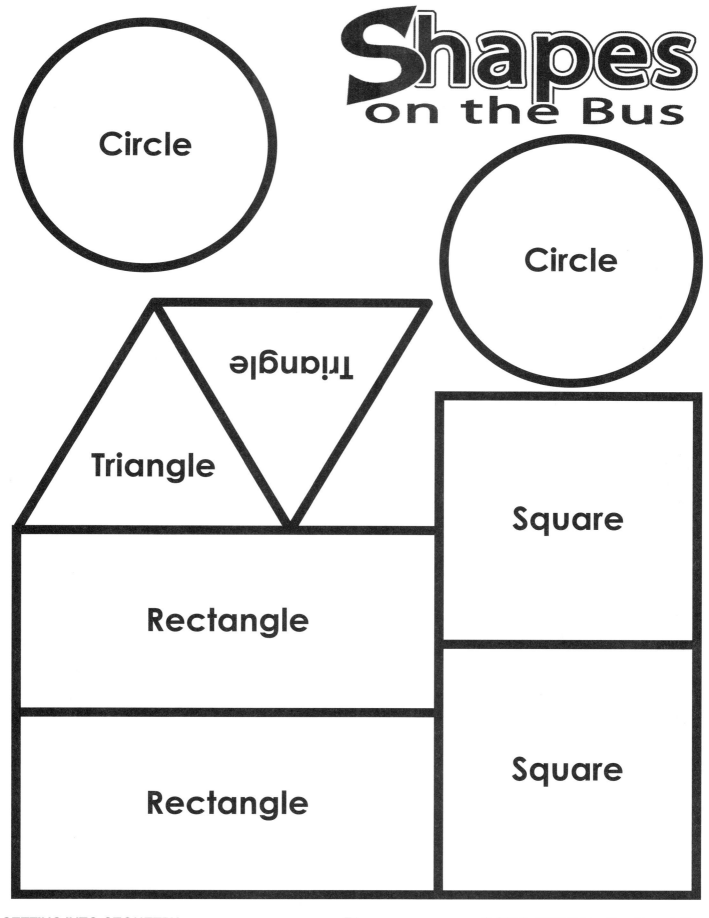

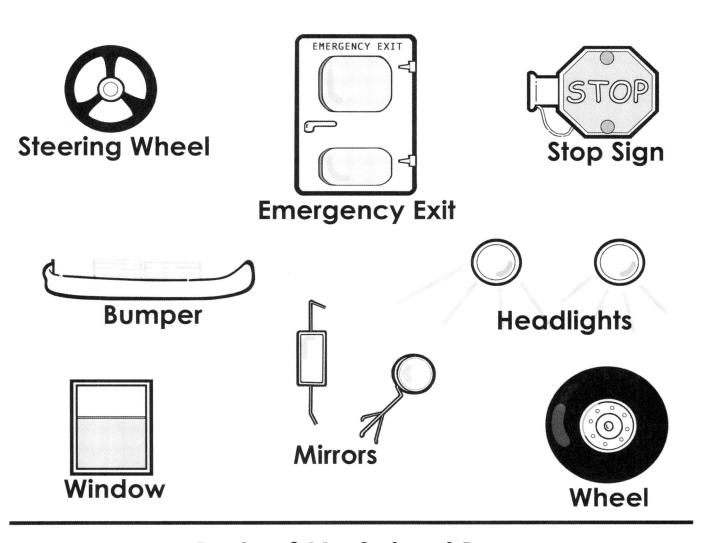

Parts of My School Bus

Shapes on the Bus—Edible Bus Recipe

Get

a piece of bread.

Spread

with cream cheese.

Put

3 square crackers on for the windows.

Add

2 round crackers for wheels.

Add

round red candies for the lights.

Eat.
M-m-m-m.

Shape World

Topic
2-D shapes

Key Question
What geometric shapes can you identify in photographs?

Learning Goals
Students will:
- identify geometric shapes in photographs, and
- recognize that geometric shapes can be seen in the real world.

Guiding Documents
Project 2061 Benchmarks
- *Circles, squares, triangles, and other shapes can be found in things in nature and in things that people build.*
- *Numbers and shapes can be used to tell about things.*
- *Shapes such as circles, squares, and triangles can be used to describe many things that can be seen.*

NCTM Standards 2000*
- *Recognize, name, build, draw, compare, and sort two- and three-dimensional shapes*
- *Describe attributes and parts of two- and three-dimensional shapes*
- *Recognize and represent shapes from different perspectives*
- *Recognize geometric shapes and structures in the environment and specify their location*

Math
Geometry
    2-D shapes
        identification

Integrated Processes
Observing
Comparing and contrasting
Identifying

Materials
Part One:
    computer with projection device
    photographs (see *Management 1*)

Part Two, Optional:
    digital camera
    printer
    white construction paper

Background Information
Shapes are present in the real world both in nature and in manufactured objects. Our buildings have rectangular doors and windows. Triangular roofs top our houses. Round wheels can be found on our cars and bicycles. Students may never have taken the time to recognize the presence of these shapes in their daily lives. This activity gives them the opportunity to identify a variety of shapes that can be seen in the real world using photographs.

Management
1. The CD that accompanies this book contains a variety of photographs that show examples of shapes in the real world. You may also use photographs that you find on the Internet to supplement these photos.
2. In order to display the photographs, you will need a projection device that is connected to a computer.
3. *Part Two* is provided as an optional extension for those with access to a digital camera and printer. It will be easiest with multiple cameras and additional adults or cross-age tutors assisting with the photography.

Procedure
Part One
1. Ask students to identify some of the shapes that you have been studying. [circle, square, triangle, rectangle] Question them as to where they have seen examples of these shapes in the real world.
2. Tell students that you will be showing them some pictures and they will try to find as many shapes as they can in the pictures.
3. Display the photograph *EastHall.jpg*. Ask students what shapes they see in the picture. Once they have pointed out the obvious square and rectangular windows, triangular roof, and round clock, draw their attention to some of the less obvious shapes. For example, there is a triangular shadow

GETTING INTO GEOMETRY   © 2010 AIMS Education Foundation

cast by the peak of the center tower on the portion of the roof in the top left side of the picture. There are spaces in the railing around the top of the center tower that form long, thin rectangles.
4. Select additional photographs from those provided (or find your own online) and repeat the process of having students identify the shapes they see. Be sure to point out the things that students may not immediately notice, like negative space or shadows.

Part Two, Optional
1. Tell students that they will now have the opportunity to find locations for their own pictures like the ones they have been viewing.
2. Take the class outdoors and allow them some time to explore different areas around the school for locations that have shapes that are easy to see.
3. Inform students that they need to find a location that they would like to have photographed and to remember where that location is.
4. If additional adults and cameras are available, have students get into groups and take turns showing their locations with the adults supervising as they take a picture. Otherwise, this can be done as a whole class or individually at a later time. Keep track of the order of students so that there will be no confusion as to which picture belongs to which student.
5. Give each student a print of his or her photograph mounted on a larger sheet of white paper. Have the students write their names on the papers and make a record of the shapes they see in the picture by drawing and/or writing the names of those shapes.
6. Display the photographs on a bulletin board or in a hallway or other common area, or use them to make a classroom book of shapes in the real world.

Connecting Learning
Part One
1. Where can you find examples of circles in the real world? …squares? …rectangles? …triangles?
2. Which shapes did you see most often in the pictures we looked at?
3. Do you think these shapes are the most common in the real world? Why or why not?
4. What are some things that were not pictured that have lots of shapes?

Part Two
1. What shapes did you see as you looked around the school?
2. What does your picture show?
3. Which shapes can be seen in your picture?
4. Did you pick the same location for your picture as anyone else? What might be some reasons for this?
5. What else would you like to take a picture of to show lots of shapes?

* Reprinted with permission from *Principles and Standards for School Mathematics*, 2000 by the National Council of Teachers of Mathematics. All rights reserved.

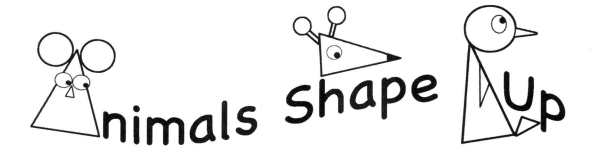

Animals Shape Up

Topic
2-D shapes

Key Question
How can you draw animals using geometric shapes?

Learning Goals
Students will:
- draw animals using shapes according to the teacher's directions, and
- create their own animals using geometric shapes.

Guiding Documents
Project 2061 Benchmarks
- *Numbers and shapes can be used to tell about things.*
- *Shapes such as circles, squares, and triangles can be used to describe many things that can be seen.*

NCTM Standards 2000*
- *Recognize, name, build, draw, compare, and sort two- and three-dimensional shapes*
- *Describe attributes and parts of two- and three-dimensional shapes*
- *Recognize and represent shapes from different perspectives*

Math
Geometry
 2-D shapes

Integrated Processes
Observing
Comparing and contrasting
Recording
Interpreting

Materials
White paper
Crayons or colored pencils
Transparency pens (see *Management 4*)

Background Information
While young students may be able to identify a variety of two-dimensional shapes, they often have little practice drawing these shapes for themselves. This activity gives them the opportunity to draw circles, squares, rectangles, and triangles while following specific directions. The opportunity is also present for reinforcing positional words such as above, below, on either side, inside, etc.

As students compare their animals to those drawn by their classmates, they will also recognize the variety that can be present among shapes of the same kind. Not all triangles are the same shape, circles can be different sizes, rectangles can be short and fat or long and skinny, etc.

Management
1. This activity is intended to follow activities in which students have learned to identify two-dimensional shapes and their properties.
2. Each student will need one sheet of paper for each animal being drawn.
3. The suggested animals use only triangles, circles, rectangles, and squares. Feel free to incorporate additional shapes as appropriate for your students.
4. If you do not have an overhead projector or document camera, you will need to adjust the presentation of the activity based on the technology you have available.

Procedure
1. Tell students that they are going to create some artwork based on your instructions. Explain that you will give them specific directions that they will need to follow and that at the end, everyone will compare their drawings to see how they are alike and how they are different.
2. Distribute a sheet of paper and crayons or colored pencils to each student.
3. Tell the class that the first picture they will be drawing will be that of a mouse. Direct them to draw a large triangle on their paper. As you describe the process, demonstrate using the projection system you have.

GETTING INTO GEOMETRY © 2010 AIMS Education Foundation

4. Tell the students to put two medium-sized circles at the top of the triangle, one on each edge. Have them draw two smaller circles inside the triangle to make eyes. Show them how to add black circles to make pupils. Direct them to draw a tiny triangle between the eyes to make a nose.

5. Have students share their illustrations. Discuss similarities and differences.
6. Distribute another sheet of paper, or use the back sides of the ones the students have.
7. Tell students that you will give instructions for making a bird, but this time you are not going to show them your drawing as you go through the steps.
8. Instruct them to draw a large circle for the body of the bird in the lower part of their papers. As they are drawing, draw a circle on the projector, but do not turn on the light.
9. Instruct students to draw a second, smaller circle on top of the first for the head of the bird and to put two circles inside this smaller circle for eyes. Do this yourself on the projector.
10. Have students draw a triangle under the eyes for the bird's beak. Instruct them to draw two triangles for wings, one on either side of the bird's body.
11. When you have finished your drawing and the class is also done, turn on the projector so that students can see your bird. Ask, "Whose bird looks like mine?"

12. Allow time for students to share their birds, comparing them with their classmates' and to yours. Discuss any things that might be incorrect, such as a square being drawn instead of a circle, as well as the differences that should be expected—not all triangles look the same, so everybody's wings and beak will be a little different; the circles are different sizes from drawing to drawing; the positions of the eyes, wings, and beak are different; etc.
13. Give students another sheet of paper and challenge them to create their own animals using only geometric shapes. When they are done, have them trade papers and identify the shapes in another person's drawing.
14. Display the animal drawings on a wall of geometric art.

Connecting Learning
1. How many circles did we use to make the mouse? How were the circles different?
2. What other shape(s) did we use? [triangle] Were the triangles the same? Explain.
3. How did your bird compare to the birds drawn by others? …by me?
4. Do all triangles look the same? Explain. [No. Some triangles have edges that are all the same, other triangles have edges that are all different, some are tall and skinny, others are short and fat, etc.]
5. How do you know something is a triangle? [It has three edges and three corners.]
6. Do all rectangles look the same? …circles? …squares? Explain. [All circles (and squares) look the same, but they can be different sizes. Rectangles can be different shapes. Some are long and skinny, others are tall and wide, etc.]
7. How do you know that a shape is a rectangle? …circle? …square? [Rectangles have four corners and four edges. The opposite edges of a rectangle are the same length. Circles are round. Squares have four corners and four edges that are all the same length.]
8. What shapes did you use in the animal you drew?
9. What shapes did you find in the animal that someone else drew?

Solutions

The following animals are shown to give you some ideas of additional pictures that students can be instructed to draw. Feel free to create your own or to modify these to incorporate other shapes appropriate for your students.

* Reprinted with permission from *Principles and Standards for School Mathematics*, 2000 by the National Council of Teachers of Mathematics. All rights reserved.

Puzzle Makers

Topic
Composing and decomposing shapes

Key Question
What hidden shapes can you find in a square?

Learning Goals
Students will:
- decompose squares, rectangles, triangles, and circles;
- identify the smaller shapes into which these shapes can be divided; and
- put smaller shapes together to form new shapes.

Guiding Document
*NCTM Standards 2000**
- *Investigate and predict the results of putting together and taking apart two- and three-dimensional shapes*
- *Build new mathmatical knowledge through problem solving*

Math
Geometry
  composing shapes
  decomposing shapes

Integrated Processes
Observing
Comparing and contrasting
Identifying

Materials
Scissors
Paper shapes (see *Management 1*)
Letter-size envelopes, two per student
Light colored construction paper, 9" x 12"

Background Information
   Often an emphasis is placed on the concept of shape composition—creating new shapes by putting smaller shapes together. This activity provides an opportunity for students to look at shape decomposition—breaking up larger shapes into their smaller components. While an activity on decomposition may seem redundant given all of the time spent on composition, it is important for primary students to have experiences that require them to break shapes down as well as build them up. This helps to reinforce properties and characteristics of shapes in their minds and facilitates their understanding that the shapes into which triangles, rectangles, squares, etc., can be broken down are the same shapes that can be used to recreate them.

Management
1. Make enough copies of the shapes provided so that each student can have two shapes. One shape will be used in *Part One*, and one shape will be used in *Part Two*. (You may also choose to cut shapes from a die cut machine, but these shapes will be smaller than the ones provided.) The shapes should be copied on card stock and laminated or cut out of foam for extended use.
2. For *Part One*, make enough puzzles for one per student. To make a puzzle, cut a shape into four smaller shapes, put the pieces into an envelope, and attach the label showing a picture of the shape on the outside of the envelope. Students will arrange the pieces into the shape shown on the envelope. Some suggested cuts are shown here.

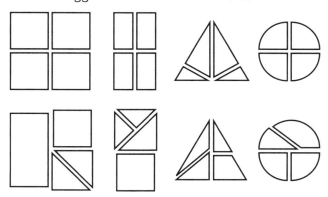

3. Leave the additional shapes uncut for *Part Two*. Students will use them to make their own puzzles.
4. While it is not the focus of this lesson, there are several opportunities to discuss additional math concepts such as symmetry and fractions.
5. If you do not have an overhead projector or a document camera, you will need to modify the procedure to suit the technology available.

GETTING INTO GEOMETRY   © 2010 AIMS Education Foundation

Procedure

Part One

1. Ask the students if they have ever put a puzzle together. Discuss how there are puzzles with different numbers of pieces and that the more pieces a puzzle has, the more difficult it usually is to solve.
2. Show the students a large triangle and ask them to identify the shape. Tell them that you are going to make a two-piece puzzle out of the triangle. Cut the triangle in two from one corner to the opposite side and project the two pieces using an overhead projector or document camera. Have the students identify the two new shapes. [triangles]

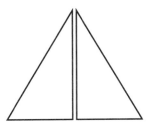

3. Invite a student to come forward and put the puzzle together so that it makes the original triangle. Discuss how a big triangle can be divided into two smaller triangles.
4. Tell the class that you are now going to cut each of the two triangles so that you can make a four-piece puzzle. Draw student's attention to one of the triangles on the projector and ask for suggestions as to how you should cut the shape to make two smaller pieces. Cut the piece into two triangles by cutting from a corner to the opposite side and ask the students to identify the new shapes. [triangles]

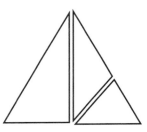

5. Cut the other triangle in the same way and invite a student to come to the projector and put the pieces together to form the original triangle.
6. Ask the students if they would each like to have a shape puzzle to solve. Distribute the shape puzzles and explain that each puzzle envelope has a picture of their shape on the outside of the envelope and the puzzle pieces inside.
7. Allow time for students to solve several puzzles and end with a discussion about what smaller shapes the larger shapes were divided into.

Part Two

1. Review what was previously done in *Part One* of this lesson.
2. Hand out an uncut shape, scissors, an envelope, and a light colored piece of construction paper to each student. Ask students to identify their shapes and trace them on the construction paper. Explain that this tracing will be the work mat others will use to solve the puzzle so they know the original shape they are trying to recreate.
3. Explain to students that they are going to look for the hidden shapes within these shapes. Show students how to find different hidden shapes by folding the shapes one time in a variety of ways. For example, fold from corner to corner, in half horizontally, in half vertically, etc.
4. Have the students fold their shapes one time and cut along the fold line. Ask several students to identify the two new shapes.
5. Ask the students to put their two-piece puzzles together to make the original shapes.
6. Allow students to make at least one more cut to one of their puzzle pieces, then place the pieces in the envelope.
7. Have students exchange puzzles and solve those created by their classmates.
8. End with a discussion about what smaller shapes they recognized in the puzzles and how the puzzles were alike and different.

Connecting Learning

1. Into what shapes did we divide our triangles? …squares? …circles? …rectangles?
2. Which puzzles were easiest? Why?
3. How did you decided what puzzle pieces (shapes) you would make from your original shape?
4. Why do you think we used mats to put our puzzles together?
5. Were the puzzles that started with the same shape all the same? Why or why not? Were any of the pieces similar?

* Reprinted with permission from *Principles and Standards for School Mathematics,* 2000 by the National Council of Teachers of Mathematics. All rights reserved.

Puzzle Makers

Make enough copies of these shape pages so that each student can have two shapes. One shape will be used in *Part One*, and the other will be used in *Part Two*.

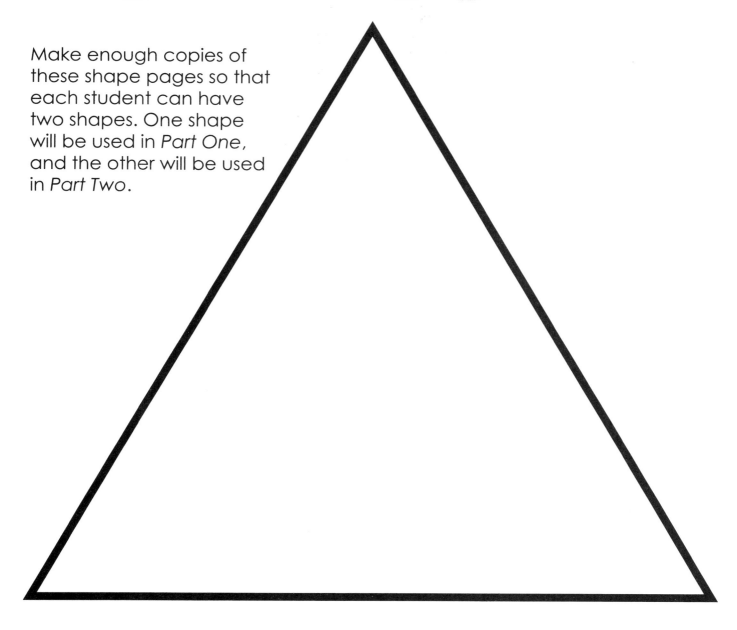

Tape this triangle label to the outside of each envelope containing a triangle puzzle for *Part One*. →

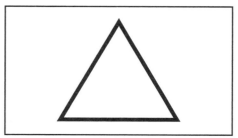

Puzzle Makers

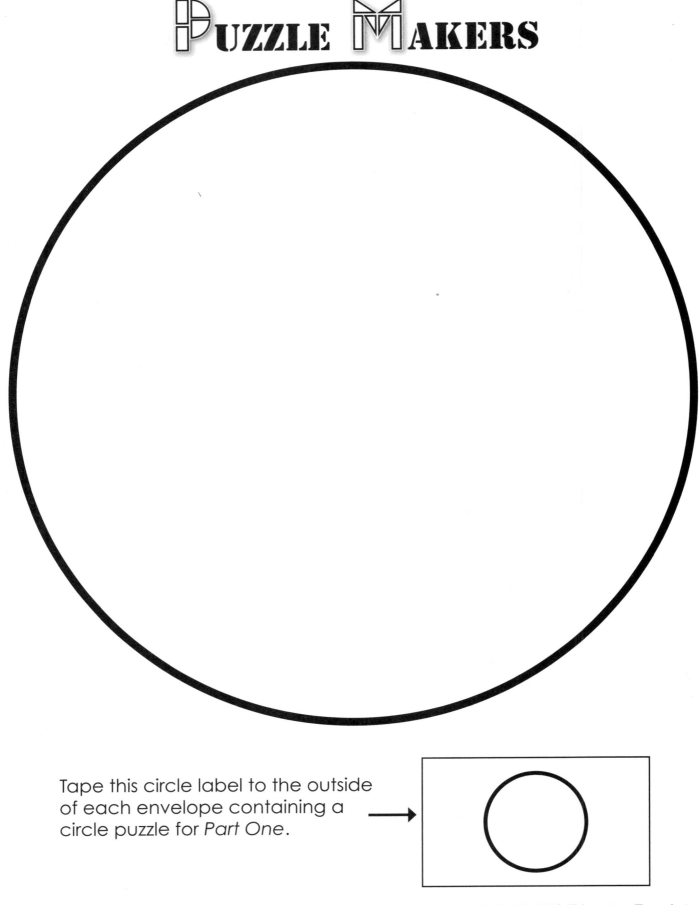

Tape this circle label to the outside of each envelope containing a circle puzzle for *Part One*.

Puzzle Makers

Tape this square label to the outside of each envelope containing a square puzzle for *Part One*. →

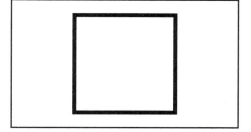

Puzzle Makers

← Tape one of these rectangle labels to the outside of each envelope containing a rectangle puzzle for *Part One*. →

GETTING INTO GEOMETRY 86 © 2010 AIMS Education Foundation

Piece-by-Piece Pictures

Topic
Composing shapes

Key Question
Which tangram pieces do I need to use to make the pictures?

Learning Goal
Students will use tangrams to compose geometric shapes into pictures.

Guiding Document
NCTM Standards 2000*
- *Investigate and predict the results of putting together and taking apart two- and three-dimensional shapes*
- *Build new mathmatical knowledge through problem solving*

Math
Geometry
    composing shapes
Problem solving

Integrated Processes
Observing
Comparing and contrasting

Materials
Student pages
Tangrams (see *Management 1*)

Background Information
   This activity invites students to superimpose tangrams over pictures in which the tangram shapes are illustrated. In filling the pictures, young learners will need to attend to the shape and its size and orientation. Care should be taken to help children understand that the tangram must fill the outline. Informal uses of transformations (slides, flips, and turns) are integral to the positioning of the tangram pieces.

Management
1. Foam tangrams are available from AIMS (item number 4182). Tangrams can also be made by cutting them from craft foam using a tangram die. Foam tangram pieces are ideal for young learners; the foam pieces are easy to pick up and they are quiet. If foam options are not possible, copy the tangram patterns provided on card stock. Each student needs one set of tangram pieces.
2. If students have not been introduced to tangrams, spend some time letting them identify and explore the shapes.
3. Cut apart the tangram picture cards along the bold lines. Distribute the pictures one at a time. They are sequenced from least difficult to most difficult.
4. It may be necessary to introduce the students to the parallelogram. One way to do this is to find how it is like a rectangle and how it is different.

Procedure
1. Distribute a set of tangram shapes to each student. Invite students to count the pieces. [7] Ask them to describe the pieces. [Some have three edges and three corners. Some have four edges and four corners. There are two large triangles, one medium triangle, two small triangles, one square, and one parallelogram. Etc.]
2. Tell students you are going to give them a page that has pictures of the tangram pieces on it. They are to match each tangram piece to its picture by placing the piece on top of the picture. Tell them to make sure that each piece fits exactly—not too big, not too small.
3. Once all children have matched the tangram pieces to the pictures, inform them that they are going to combine the pieces to make shapes and pictures. Distribute the first card. Ask students to identify what is on the card. [a square] Have them find the pieces that make up the square and put them in place on the card. Check their results to make sure they have correctly aligned and oriented the pieces.
4. Challenge students to find another tangram piece that could cover the shape. [the square] Discuss how the two small triangles can make the square shape.
5. Repeat this process for the first six cards to help reinforce the relationships among the tangram pieces.
6. Tell students that they will now be using their tangrams to make some pictures. Distribute the seventh card and ask students what they think it looks like. [a tree] Ask them what pieces make up the tree. [a large triangle, the medium triangle, and a square]
7. Distribute the other pictures, checking each child's results.
8. Conclude with a discussion about how they used the tangram pieces to compose pictures.

GETTING INTO GEOMETRY © 2010 AIMS Education Foundation

Connecting Learning

1. Which picture did you like best? What pieces were used to make it?
2. What shapes can two small triangles make? [square, medium triangle, parallelogram]
3. What shapes can you use to make the large triangle? [the square and the two small triangles, the medium triangle and the two small triangles, or the parallelogram and the two small triangles]
4. How many pieces were in the little tree? [3] Name the pieces. [the square, the medium triangle, and a large triangle]
5. How many pieces were in the big tree? [4] Name the pieces. [the square, a large triangle, the medium triangle, and a small triangle]
6. What pieces were the same in the small boat and the large boat? [small triangle, large triangle, parallelogram]
7. What piece was different in the large boat? [a small triangle]
8. Which picture used all the pieces? [the large square]
9. What picture can you make using the tangram pieces? What pieces will you combine?

* Reprinted with permission from *Principles and Standards for School Mathematics*, 2000 by the National Council of Teachers of Mathematics. All rights reserved.

Piece-by-Piece Pictures

If you do not have tangram puzzles, copy this page on card stock and cut apart the puzzle pieces. Each student will need one set.

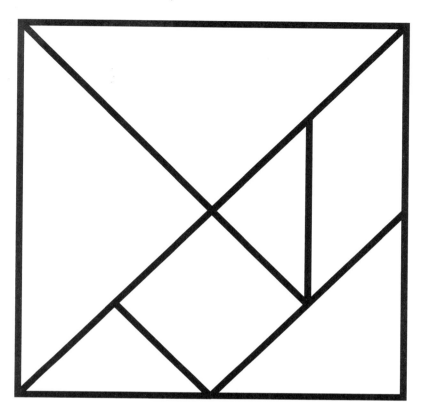

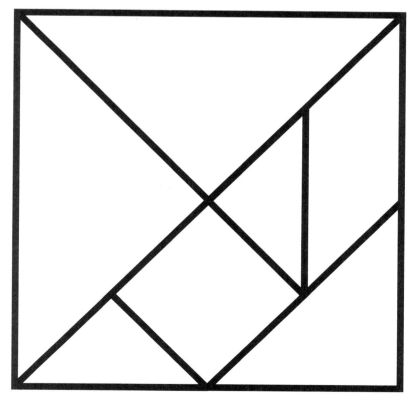

GETTING INTO GEOMETRY © 2010 AIMS Education Foundation

Piece-by-Piece Pictures

Match your pieces to the shapes.

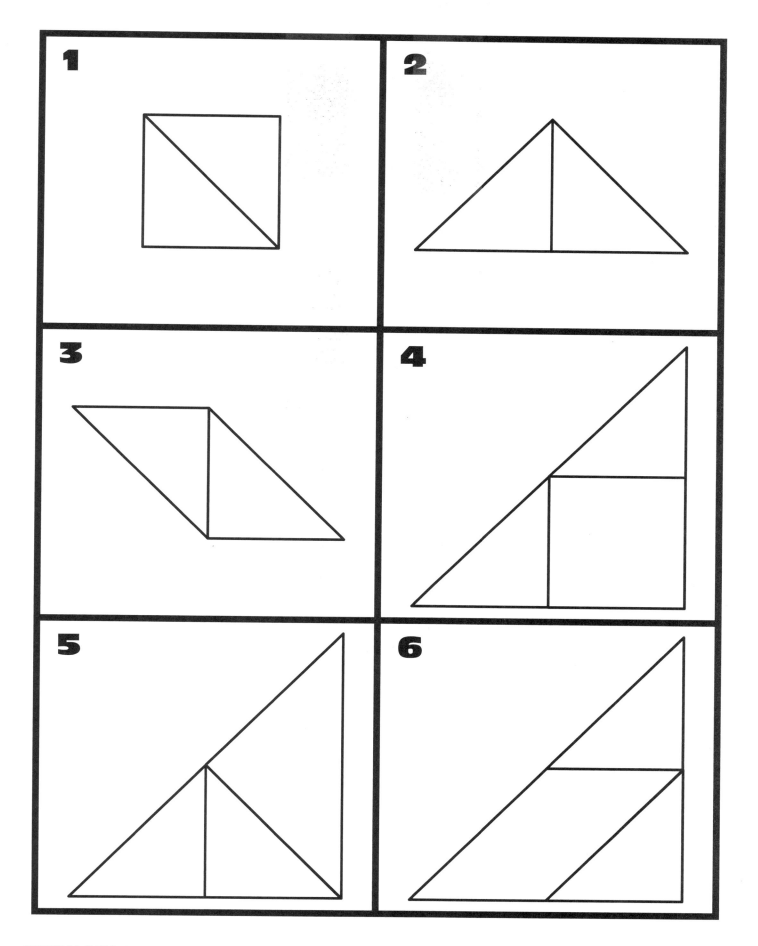

Piece-by-Piece Pictures

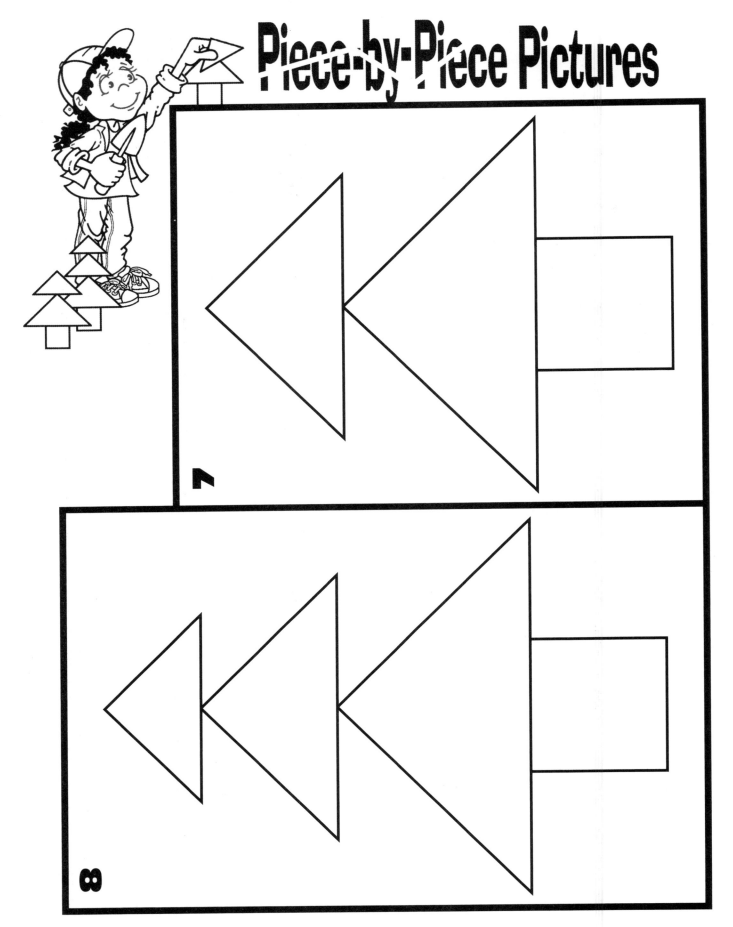

Piece-by-Piece Pictures

Piece-by-Piece Pictures

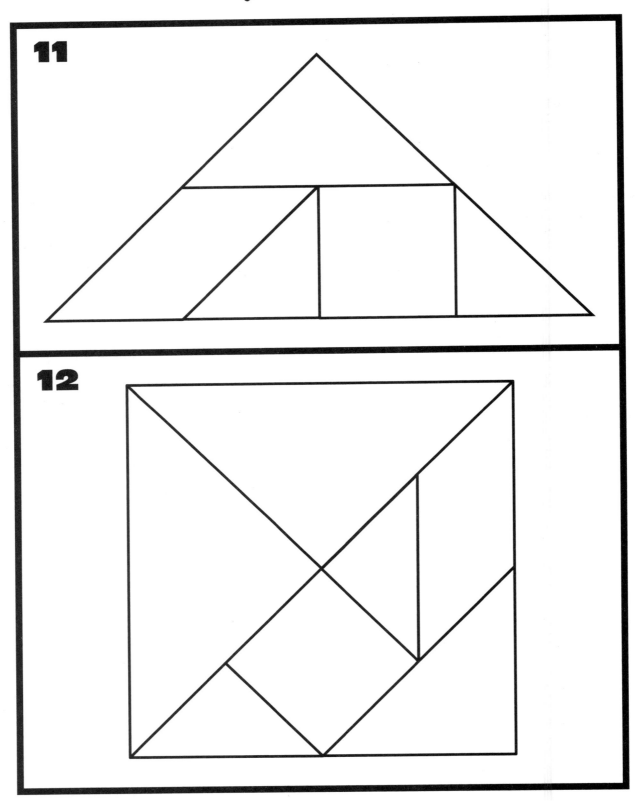

GETTING INTO GEOMETRY © 2010 AIMS Education Foundation

Picture Perfect

Topic
Composing shapes

Key Question
Which tangram pieces do I need to use to fill the outlines of the pictures?

Learning Goal
Students will use tangrams to fill outlines of pictures.

Guiding Document
NCTM Standards 2000*
- *Investigate and predict the results of putting together and taking apart two- and three-dimensional shapes*
- *Build new mathematical knowledge through problem solving*

Math
Geometry
    composing shapes
Problem solving

Integrated Processes
Observing
Comparing and contrasting
Recording

Materials
Student pages
Tangrams (see *Management 1*)

Background Information
    This activity invites students to compose geometric shapes into pictures. Care should be taken to help children understand that the tangram must fill the outline. Students will use the problem-solving strategies of guess and check and logical thinking. Informal uses of transformations (slides, flips, and turns) are integral to the positioning of the tangram pieces.

Management
1. Foam tangrams are available from AIMS (item number 4182). Tangrams can also be made by cutting them from craft foam using a tangram die. Foam tangram pieces are ideal for young learners; the foam pieces are easy to pick up and they are quiet. If foam options are not possible, copy the tangram patterns provided on card stock. Each student needs one set of tangram pieces.
2. This activity is intended to follow *Piece-by-Piece Pictures*.
3. Students should work in pairs with each student making his or her own record of results.

Procedure
1. Distribute a set of tangram shapes to each student. Tell students that their challenge will be to fill outlines of pictures with some of the tangram pieces. Explain that there may be more than one way to solve each problem, but that they will never need to use all seven of their tangrams for one picture.
2. Display the picture of the house as a class example. Encourage students to give suggestions as to which pieces to try. [The large triangle fits perfectly to form the roof of the house.] Guide students to understand that they now must fit three pieces into the lower part of the house.
3. Request volunteers to assist in filling the outline. When the outline has been completely filled, show students how to trace around one tangram piece at a time so there is a record on the page.
4. Distribute the first student page. Direct students to fill in the outline of the house with their tangrams. Assist as needed as they trace around the tangrams to make a record.
5. Have students work together to fill in the outline of the mountains and to trace their solutions, recording how many pieces they used.
6. Once students understand the task, distribute the other student pages one at a time, checking each child's results.
7. Conclude with a discussion about how they used the tangram pieces to compose pictures. Have students compare the different ways they were able to fill the pictures and see if any additional solutions can be found.

Connecting Learning
1. How did we use tangrams to make pictures? [We had to put the pieces together.]
2. Which picture did you like best?
3. What pieces were used to make it?
4. What picture was the most difficult to make? What made it so difficult?
5. How did you decide which tangrams to try?
6. Do you think there are any other pictures that could be made with our tangram pieces? Explain your answer.

GETTING INTO GEOMETRY

Solutions

There are multiple ways to solve all of the challenges that use both different arrangements of pieces and/or different numbers of pieces. At least one solution for each card is given here.

House

Mountains

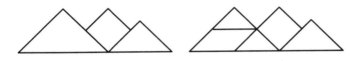

Barn

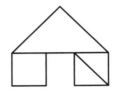

Fighter Plane

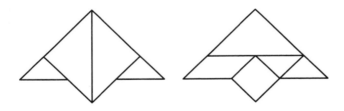

Tulip

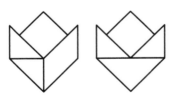

Sailboat

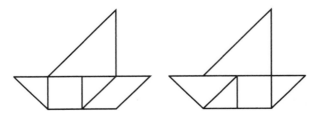

Crown

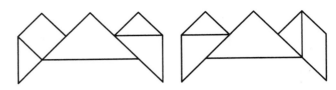

Square

Rectangle

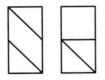

* Reprinted with permission from *Principles and Standards for School Mathematics*, 2000 by the National Council of Teachers of Mathematics. All rights reserved.

Picture  Perfect

If you do not have tangram puzzles, copy this page on card stock and cut apart the puzzle pieces. Each student will need one set.

GETTING INTO GEOMETRY © 2010 AIMS Education Foundation

Picture Perfect

1 House

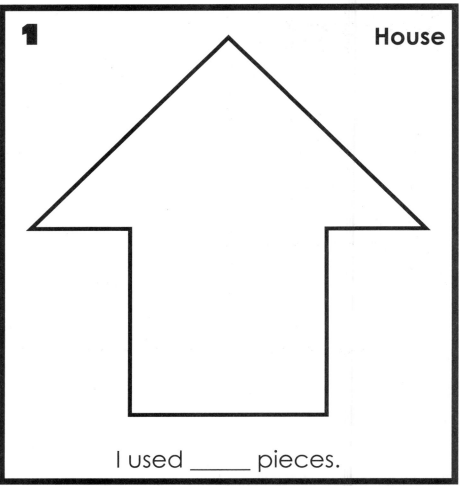

I used ____ pieces.

2 Mountains

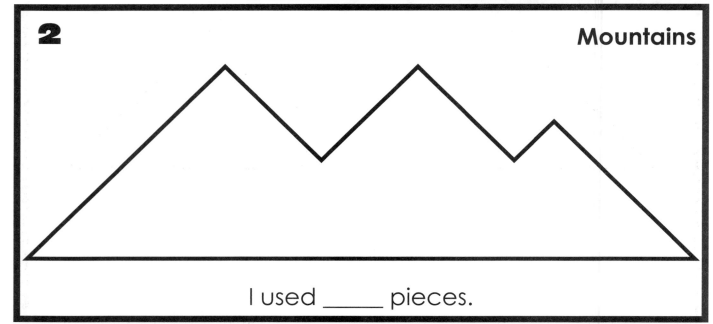

I used ____ pieces.

Picture Perfect

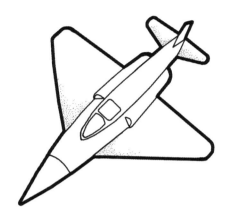

3   Barn

I used _____ pieces.

4   Fighter Plane

I used _____ pieces.

Picture Perfect

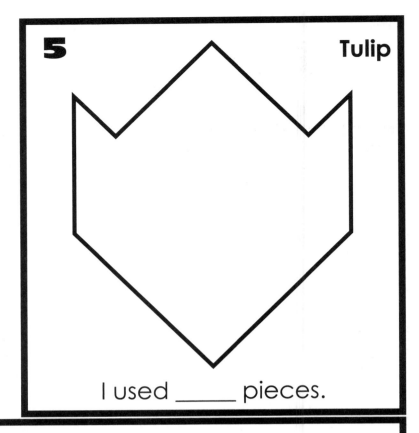

5 **Tulip**

I used ____ pieces.

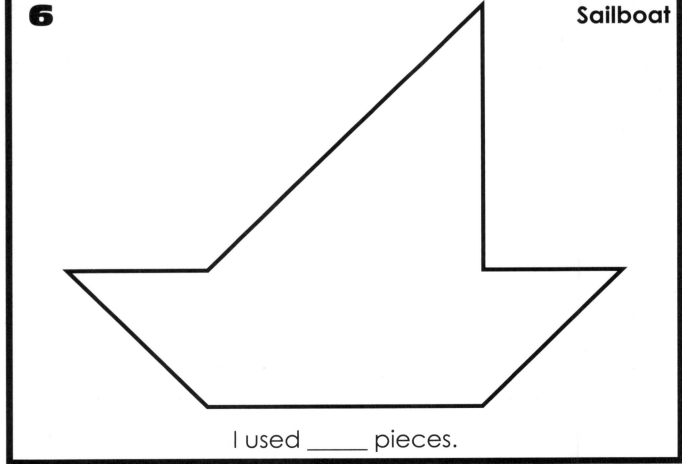

6 **Sailboat**

I used ____ pieces.

Picture Perfect

7 Crown

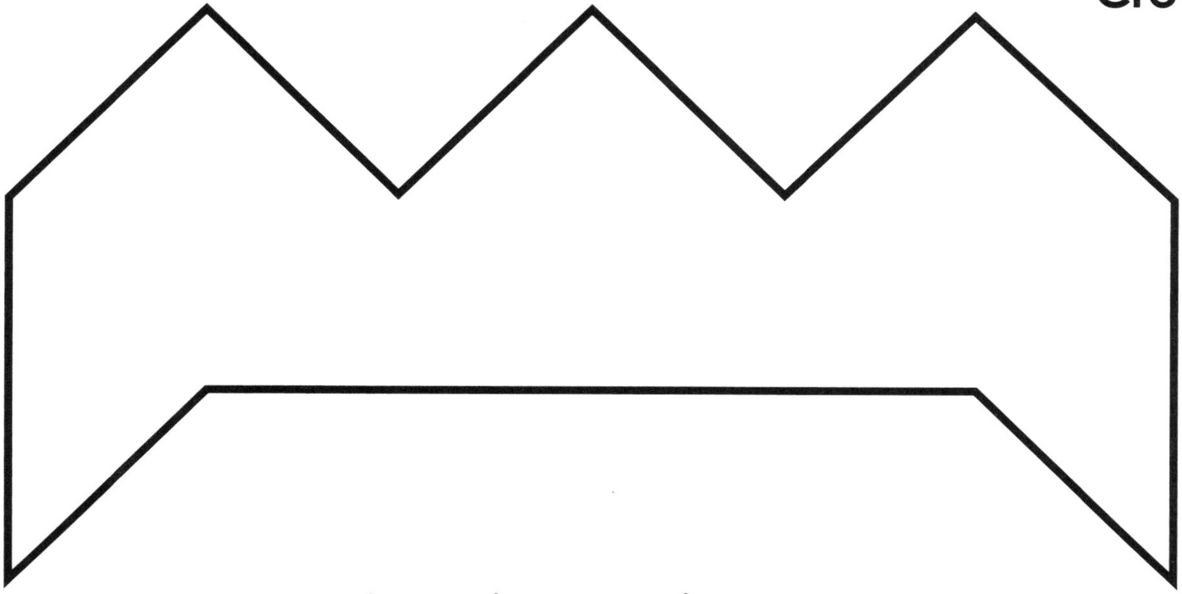

I used _____ pieces.

8 Square

I used _____ pieces.

9 Rectangle

I used _____ pieces.

GETTING INTO GEOMETRY © 2010 AIMS Education Foundation

Topic
Composing shapes

Key Question
Which tangram pieces do I need to use to make the pictures?

Learning Goal
Students will use small diagrams and tangram pieces to make pictures.

Guiding Document
NCTM Standards 2000*
- *Investigate and predict the results of putting together and taking apart two- and three-dimensional shapes*
- *Build new mathematical knowledge through problem solving*

Math
Geometry
   composing shapes
Problem solving

Integrated Processes
Observing
Comparing and contrasting

Materials
Student page
Tangrams (see *Management 1*)

Background Information
This activity invites students to compose geometric shapes into pictures. Young learners will look at reduced-sized illustrations and translate them into compositions made with their tangram pieces. To assist them in doing this, one shape in the enlarged picture has been placed to give them a starting point. They will need to consider the shape and its size and orientation. Care should be taken to help children carefully attend to the relative size of the pieces. In each case where triangles are used, students will have to determine which of the three sizes belongs. Informal uses of transformations (slides, flips, and turns) are integral to the positioning of the tangram pieces.

Management
1. Foam tangrams are available from AIMS (item number 4182). Tangrams can also be made by cutting them from craft foam using a tangram die. Foam tangram pieces are ideal for young learners; the foam pieces are easy to pick up and they are quiet. If foam options are not possible, copy the tangram patterns provided on card stock. Each student needs one set of tangram pieces.
2. This is not intended to be an introductory experience with tangrams. It is recommended that students do the activities *Piece-by-Piece Pictures* and *Picture Perfect* prior to this one.
3. Cut apart the tangram picture cards on the first two student pages. Distribute the pictures one at a time. They are sequenced from least difficult to most difficult.

Procedure
1. Distribute a set of tangram shapes to each student along with the first card with the diagram of the arrow. Ask students what shapes make up the arrow. [two small triangles and a medium triangle]
2. Ask them if they can place the tangram pieces on top of the small picture and make them fit. [No.] Why? [The picture is too small.]
3. Encourage students to make suggestions as to how they could use their tangrams to make the picture. [Place the tangrams on the card next to the picture.]
4. Help students make the arrow, assisting as needed.
5. Once students understand the task, distribute the other pictures one at a time, checking each child's results.
6. Conclude with a discussion about how they used the tangram pieces to compose pictures.

Connecting Learning

1. How did we use tangrams to make pictures? [We had to put the pieces together.]
2. Which picture did you like best?
3. What pieces were used to make it?
4. What picture was the most difficult to make? [probably the rabbit] What made it so difficult? [There were a lot of pieces.]
5. What pieces made the swan's head? [the small triangle and the square]
6. What pieces made the rabbit's ears? [the small triangle and the parallelogram]
7. How are the triangles alike? [They all have three sides and three corners.]
8. How are they different? [They are different sizes.]
9. Which picture used all the pieces? [the rabbit]

* Reprinted with permission from *Principles and Standards for School Mathematics*, 2000 by the National Council of Teachers of Mathematics. All rights reserved.

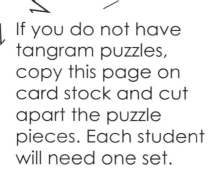

If you do not have tangram puzzles, copy this page on card stock and cut apart the puzzle pieces. Each student will need one set.

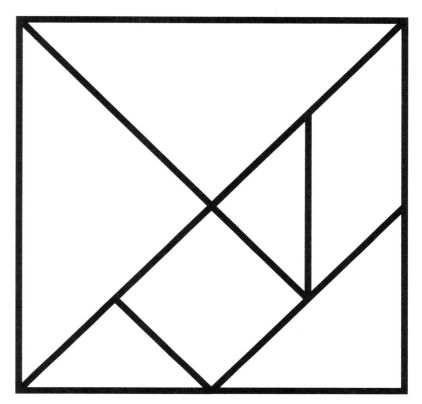

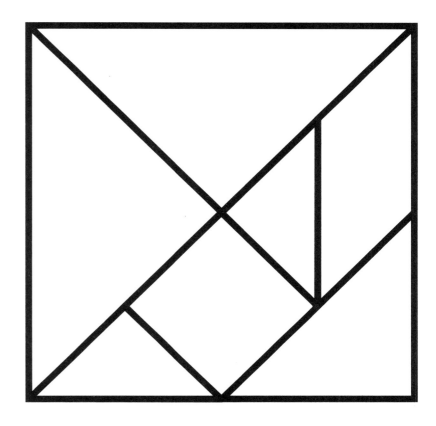

GETTING INTO GEOMETRY © 2010 AIMS Education Foundation

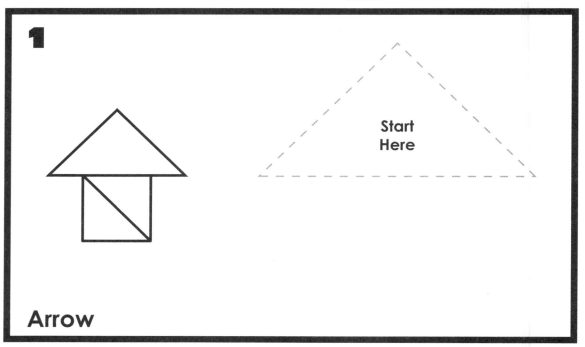

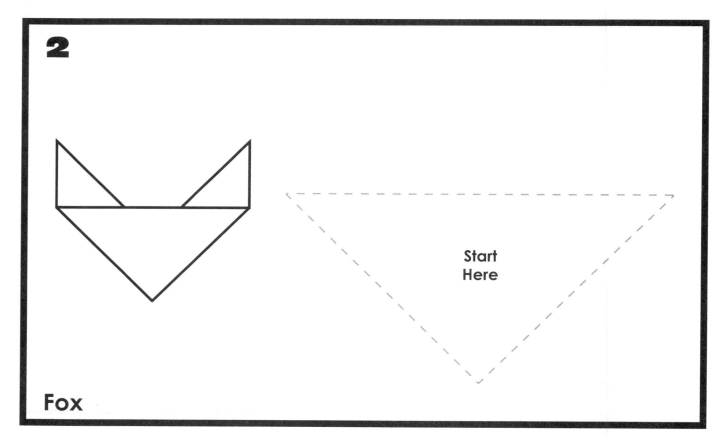

GETTING INTO GEOMETRY 106 © 2010 AIMS Education Foundation

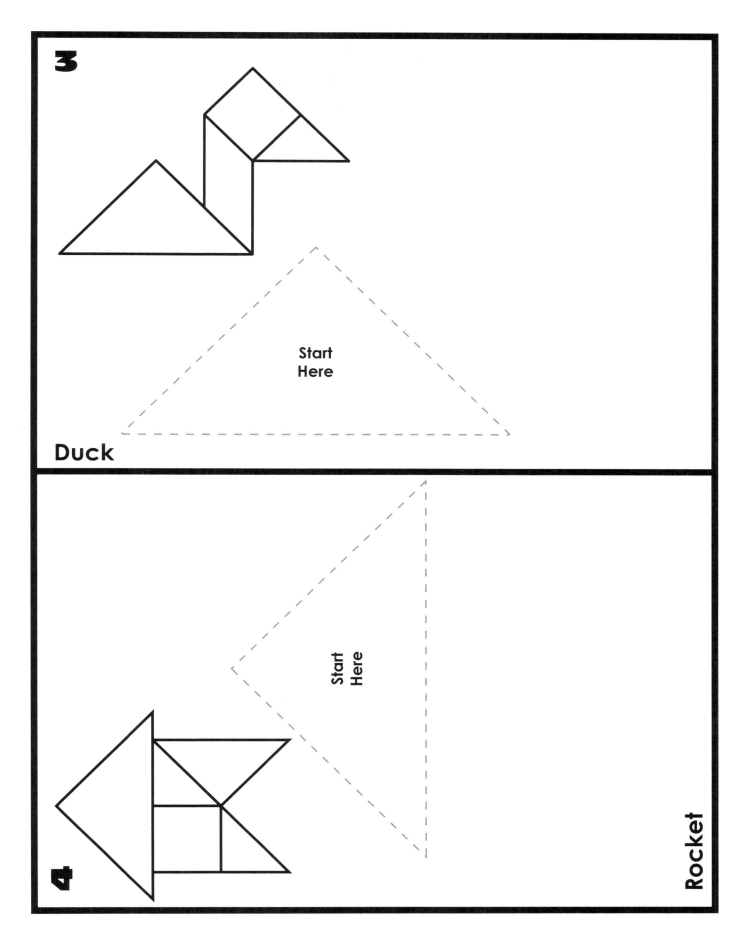

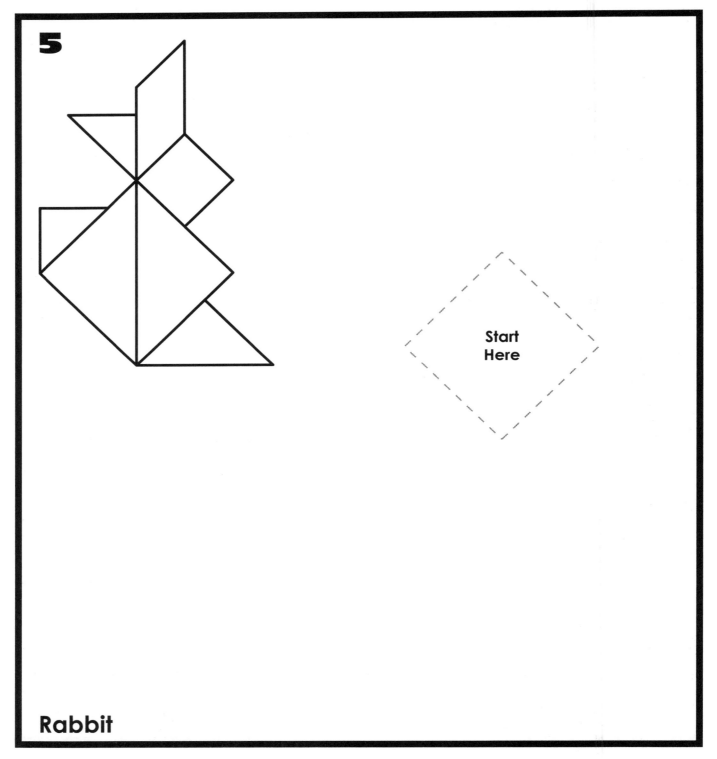

5

Rabbit

Topic
Symmetry

Key Question
What makes a shape symmetric?

Learning Goals
Students will:
• define symmetry,
• fold paper shapes to find the line(s) of symmetry for each shape, and
• cut shapes of their own and determine if those shapes are symmetric.

Guiding Documents
Project 2061 Benchmark
• Many objects can be described in terms of simple plane figures and solids. Shapes can be compared in terms of concepts such as parallel and perpendicular, congruence and similarity, and symmetry. Symmetry can be found by reflection, turns, or slides.

NCTM Standard 2000*
• Recognize and create shapes that have symmetry

Math
Geometry
 symmetry

Integrated Processes
Observing
Comparing and contrasting
Classifying

Materials
Shapes (see *Management 1*)
Flash cards, included
Two small boxes (see *Management 2*)
Scratch paper
Scissors

Background Information
When a shape has line symmetry, it can be divided into two halves that are mirror images of each other. The line along which it is divided is the line of symmetry. Depending on the shape, there may be multiple lines of symmetry.

When teaching symmetry to young children, the new vocabulary should be introduced through hands-on activities and games. This lesson is an introductory level activity that will give students the opportunity to fold shapes to find the line(s) of symmetry within each shape.

Management
1. Prior to teaching this lesson, copy and cut out the included set of shapes for each student: circle, square, triangle, and rectangle. If there are die cuts available of the shapes, use these to make a set of shapes for each student. Also copy and cut apart the flash cards.
2. Prepare two small boxes—one labeled *Symmetric* and one labeled *Not Symmetric*—for *Part Two* of this lesson. When the lesson is finished, the boxes and shapes can go to a center for additional practice identifying symmetric shapes.

Procedure
Part One
1. Display the shape flash cards one at a time and ask the students to identify the shapes.
2. Tell the students that some shapes can be folded in half so that both halves are exactly the same, and that those shapes are called *symmetric shapes*. Explain that the line that is made in the middle when you fold the shape is called the *line of symmetry*. Explain that not all shapes have a line of symmetry and that some shapes have more than one line of symmetry.
3. Tell the class that they are going to fold some shapes to find out if they are symmetric.
4. Give each student the set of cut out shapes.
5. Have the students find the square in their sets of shapes. Tell them to fold it in half. Demonstrate the fold before they fold. Ask them to turn the shape over to see if it is the same on both sides of the fold. Have them open the shape. Explain that if the shape is the same on both sides of the fold, then it is symmetric and the fold line is the line of symmetry.
6. Before moving on to a new shape, ask the students if they can see other ways to fold the square so that it would be the same on both sides of the fold. Challenge them to find all of the possible lines of symmetry for a square. [through the center horizontally, vertically, and along both diagonals]
7. Have the students find the circle in their set of shapes. Ask them to predict how many lines of symmetry the circle would have. Discuss why the circle may have so many lines of symmetry. Have them fold the circle several times and then move on to the rest of the shapes.

GETTING INTO GEOMETRY 109 © 2010 AIMS Education Foundation

8. Repeat folding and identifying the lines of symmetry for the rest of the shapes. When folding the triangle, question students about why it only has one line of symmetry. When folding the rectangles, compare the lines of symmetry with those of the square.
9. End with a discussion about how to identify symmetric shapes and lines of symmetry.

Part Two
1. Review the meaning of symmetry.
2. Hand each student several pieces of scratch paper and scissors and ask them to cut out some shapes of their own.
3. When students have several shapes cut out, ask them to test them by folding to see if their shapes are symmetric.
4. Show the students the two boxes you have labeled *Symmetric* or *Not Symmetric*.
5. Have students choose one of the shapes that they cut and allow them to tell the class whether it is symmetric or not and why, then ask them to place their shape in the appropriate box.
6. End with a discussion about how you can determine whether or not a shape is symmetric.

Connecting Learning
1. What does *symmetric* mean?
2. How did you decide if the shape was symmetric?
3. How would you teach someone else about things that are symmetric?
4. What shapes did we decide are symmetric?
5. Are all shapes symmetric? Explain.
6. Do some shapes have more than one line of symmetry? Explain.

* Reprinted with permission from *Principles and Standards for School Mathematics,* 2000 by the National Council of Teachers of Mathematics. All rights reserved.

Shape Symmetry
Flash Cards

GETTING INTO GEOMETRY © 2010 AIMS Education Foundation

Shape Symmetry
Triangles

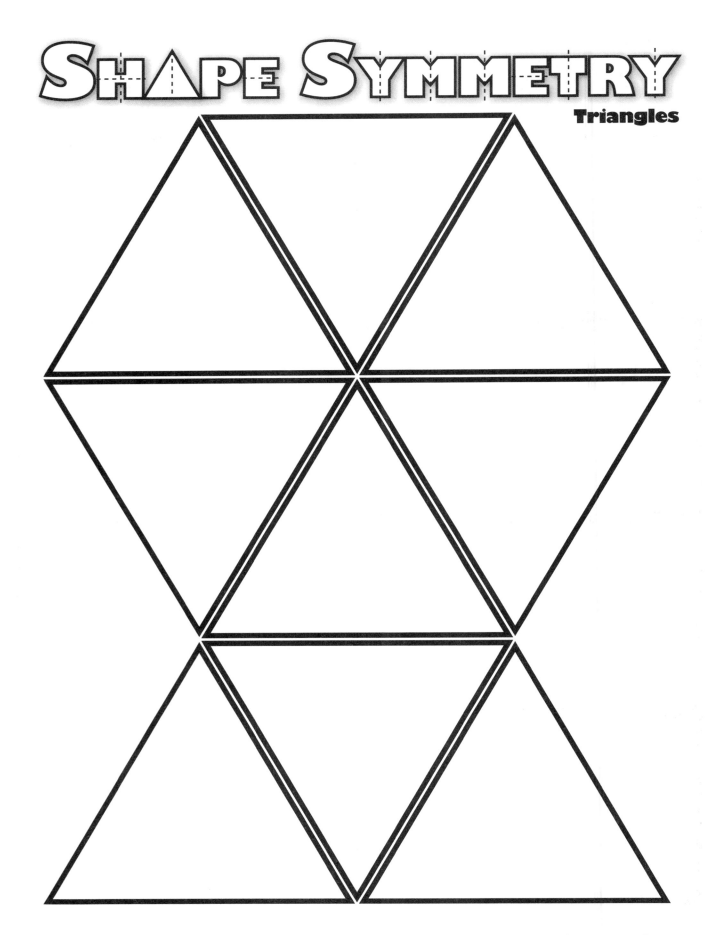

GETTING INTO GEOMETRY 114 © 2010 AIMS Education Foundation

Shape Symmetry
Rectangles

GETTING INTO GEOMETRY 115 © 2010 AIMS Education Foundation

Match Play

Topic
Symmetry

Key Question
How do you decide if a shape is symmetric?

Learning Goals
Students will:
* identify line symmetry, and
* apply their understanding of line symmetry to real-world objects.

Guiding Documents
Project 2061 Benchmark
* Many objects can be described in terms of simple plane figures and solids. Shapes can be compared in terms of concepts such as parallel and perpendicular, congruence and similarity, and symmetry. Symmetry can be found by reflection, turns, or slides.

NCTM Standards 2000*
* *Recognize and create shapes that have symmetry*
* *Recognize and apply mathematics in contexts outside of mathematics*

Math
Geometry
   symmetry

Integrated Processes
Observing
Comparing and contrasting
Communicating
Generalizing

Materials
Student pages (see *Management 1*)
Let's Fly A Kite by Stuart Murphy (see *Curriculum Correlation*)
Index cards, 3" x 5" (see *Management 2*)

Background Information
When an object with line symmetry is folded in half along the line of symmetry, the two halves will match. A square has four lines of symmetry, meaning that there are four places where it can be divided into two halves that are mirror images of each other. The lines of symmetry can be found by folding through the center horizontally, vertically, or along either diagonal. Other shapes such as irregular quadrilaterals, circles, triangles, pentagons, rectangles, etc., have zero, one, two, three, or even an infinite number of lines of symmetry.

In this activity, students will fold different pictures in half. They will see how both halves are alike when a picture is symmetric.

Management
1. For each student, make two copies of the student pages. One set will be used to cut and fold; the other set will be used for recording.
2. For *Part Two,* each student will need several index cards.
3. If the book *Let's Fly a Kite* is not available, use the *Match Play* pages and pictures from magazines to introduce students to symmetry.

Procedure
Part One
1. To introduce or review the concept of symmetry, read *Let's Fly A Kite* (see *Curriculum Correlation*) to the class.
2. As you read the book a second time, pause to discuss with students how the dashed lines in the book are dividing the pictures in half. Point out that both halves are the same. Discuss how the characters in the book, in some cases, simply folded the objects to discover the two halves.
3. Give students a copy of the first student page. Direct them to cut out each picture, leaving the frame around the pictures, and to fold the pictures in half to show two halves that are alike. Students will discover that not all pictures can be folded in half to show two halves alike. Encourage students to hold the folded papers up toward the light so they can see if the lines in the picture overlay each other. If accessible, children can hold the folded pictures up to windows to check for symmetry. (Make sure that students understand that they are trying to match the outlines of the pictures inside the rectangles that they cut out and not the rectangle itself.)

GETTING INTO GEOMETRY

4. Have students divide the pictures into two groups, those that can be divided in half with the two halves alike and those that cannot. Discuss their results.

5. Introduce or review the math vocabulary words of *symmetry* and *symmetric*. Display these words on a word wall, in a word bank, or in a pocket chart, etc. Tell students that when a picture can be divided into two halves that are alike, then that picture is symmetric or it has symmetry. Add other words such as *same, identical, twin, match*, etc., to the word wall as appropriate.

6. Discuss the pictures that cannot be folded in half with two halves alike. Explain that these pictures are *not symmetric.*

7. Have students place their fingers on one of the folded lines that divides a symmetric picture in half. Discuss how this is called a *line of symmetry.* It is a line that divides the picture in two halves that are the same. If they were to cut this picture in half, one side could lay directly over the other and appear the same.

8. To record the lines of symmetry, have students use the second copy of the first student page. Tell them to draw dashed lines along the lines of symmetry of the pictures that are symmetric.

9. Refer back to the pictures in the book and discuss how the two halves of each picture featured with dashed lines are the same. Discuss the pictures that have more than one way to divide the picture in two halves that are the same.

Part Two
1. Take the class on a *Searching for Symmetry* walk around the classroom and outdoors. Ask students to identify objects they think are symmetric. If possible, have them fold the items to check for lines of symmetry.

2. Suggest that students check art that is displayed in the classroom, leaves, flowers, etc. Have them fold a jacket in half, matching shoulder to shoulder. Ask them to fold the same jacket in half matching the shoulders to the bottom hem of the jacket.

3. On index cards, ask students to either draw a picture or to describe the objects identified on the walk. (Students will need one card for each object.) On the cards, have them record the number of lines of symmetry found in these objects. Gather the class for further discussion.

4. Discuss the pictures, objects, etc., discovered by the class. Ask them to sort the index cards according to the number of lines of symmetry found on each. Ask them to identify those with one, two, three, or more lines of symmetry. You may want students to display their cards in bar-graph fashion for ease of discussion.

5. Ask students to make true statements about their sorted sets. (For example: We found more things that had one line of symmetry than three lines of symmetry.)

6. Discuss items they explored that had no lines of symmetry.

Connecting Learning
1. How do you know if a shape is symmetric? [If you fold something in half, both halves look the same.]

2. Which shapes have more than one line of symmetry? [rectangles and squares] Which shapes had no lines of symmetry? [funny looking shapes (these may be irregular quadrilaterals, irregular polygons, etc., but young learners will more than likely not have the vocabulary to identify them)]

3. How can you check to decide if a shape is symmetric? [Fold it in half, or draw a line through the middle. If both halves are the same, then it is symmetric.]

4. What things did you find with symmetry around the classroom and outdoors?

5. How were you able to tell if these things were symmetric?

6. What other things have symmetry?

Extensions
1. Have students explore shapes such as triangles, hexagons, octagons, to find the lines of symmetry. Some of these shapes will only have one line of symmetry, others will have multiple lines of symmetry.

2. Ask students to take blank index cards and to try to draw designs or pictures or shapes with lines of symmetry.

Curriculum Correlation
Murphy, Stuart J. *Let's Fly a Kite.* HarperCollins. New York. 2000.
This book offers an introduction to horizontal and vertical symmetry cast in real-world examples.

* Reprinted with permission from *Principles and Standards for School Mathematics,* 2000 by the National Council of Teachers of Mathematics. All rights reserved.

Match Play

GETTING INTO GEOMETRY 119 © 2010 AIMS Education Foundation

Squishy Symmetry

Topic
Symmetry

Key Question
What makes something symmetric?

Learning Goal
Students will paint a picture to explore symmetry.

Guiding Documents
Project 2061 Benchmark
- *Many objects can be described in terms of simple plane figures and solids. Shapes can be compared in terms of concepts such as parallel and perpendicular, congruence and similarity, and symmetry. Symmetry can be found by reflection, turns, or slides.*

NCTM Standard 2000*
- *Recognize and create shapes that have symmetry*

Math
Geometry
   symmetry

Integrated Processes
Observing
Comparing and contrasting
Communicating

Materials
9" x 12" construction paper
Colored tempera paint (see *Management 2*)
Aluminum foil (see *Management 2*)
Plastic spoons, one per student
Paper towels

Background Information
   In learning geometry, children need to experiment and explore the concepts in a variety of ways. They learn about the properties of shapes and increase their awareness of spatial concepts by using everyday objects and looking at objects from different perspectives. Symmetry is a property of both two- and three-dimensional shapes and solids. Folding paper cutouts or using mirrors to investigate lines of symmetry are some of the ways that children use to explore shapes. It is important for students to have a variety of experiences using different materials, such as blocks, paper cutouts, and pictures from books and magazines, to look for lines of symmetry before they use symmetry abstractly.

Management
1. It is assumed that students have done the activity *Match Play* prior to this experience.
2. Cut a small piece of foil for each student to serve as a paint palette. Put a small amount of each of three or four colors of paint on each piece of foil.
3. Limit the number of times the student presses or squeezes the paper once the paint is inside. Rubbing it too many times may cause the paint to ooze out or cause the paint colors to become too intermingled.
4. Make your own picture using the technique to show students as an example. (See *Procedure 2.*)

Procedure
1. Remind the students of the experiences they have already had with symmetry. Review the concept of line symmetry and remind them of how they folded pictures in half to show they were alike on both sides.
2. Tell the students that they will be using paint to make a picture and observing if the picture is symmetric. Show students your sample symmetry picture (see *Management 4*). Explain that they will fold the paper in half and open it up. On one half of the paper, they will put four blobs of paint. The paint will be near the fold of the paper so that it will be easy to compare both sides. Tell them that they will rub the paper after they close it.
3. Hand out a sheet of constuction paper to each student. Have them fold the paper in half vertically. Assist as necessary.
4. Distribute the spoons and the pieces of foil with the paint on them. Allow students time to put the paint on their papers as they wish. Encourage them to stay away from the outer edges of the paper so that the paint doesn't squish out.

GETTING INTO GEOMETRY

5. When the students have gotten the paint on the paper, instruct them to re-fold the paper along the crease so that the paint is inside. Tell them to use their hands to squish the paint inside the paper. Two or three rubs are all that are needed. Be ready with paper towels should the paint squish out.
6. When all of the students have squished their papers, have them open the folded papers carefully. They should have designs on their papers that are symmetric.
7. Have the students set the pictures in an area where they can be left to dry. When they are dry, have students get their pictures and discuss what they see.

Connecting Learning
1. What is symmetry?
2. Was the picture you made today symmetric? [Yes.] How do you know? [Both sides are the same.]
3. Describe your picture. What are some of its interesting features? Does it remind you of anything?
4. If you folded the paper the other way, would your picture be symmetric? Explain.
5. If you wanted to teach your family how to tell if something is symmetric, what would you do? Explain.
6. Can you think of three other things that are symmetric? What are they? How do you know they are symmetric?
7. Do you think you are symmetric? Explain.
8. Look at someone sitting near you and decide if he/she is symmetric. Explain.
9. Can anyone think of something that we could do with our symmetric pictures?

* Reprinted with permission from *Principles and Standards for School Mathematics*, 2000 by the National Council of Teachers of Mathematics. All rights reserved.

The Art of Symmetry

Topic
Symmetry

Key Question
How can we make an art project that shows something in nature that has symmetry?

Learning Goal
Students will create symmetry pictures that show things in nature that have a line of symmetry.

Guiding Documents
Project 2061 Benchmark
- *Many objects can be described in terms of simple plane figures and solids. Shapes can be compared in terms of concepts such as parallel and perpendicular, congruence and similarity, and symmetry. Symmetry can be found by reflection, turns, or slides.*

NCTM Standard 2000*
- *Recognize and create shapes that have symmetry*

Math
Geometry
  symmetry

Integrated Processes
Observing
Relating

Materials
9" x 12" construction paper (see *Management 1*)
Scissors
Figure templates
File folders
Tape
Glue sticks

Background Information
When an object or shape has line symmetry, it can be divided into two halves that are mirror images of each other. If one half could be flipped and placed on top of the other half, they would match exactly. The line along which this division takes place is the line of symmetry.

This property can also be used to create figures with symmetry. Any figure can be flipped and aligned to create a new figure with symmetry, as shown in the example here.

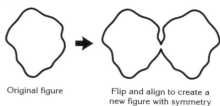

Original figure   Flip and align to create a new figure with symmetry

It is this idea that students will be using in this activity. They will cut figures from half sheets of paper. Rather than cutting the figure twice and flipping one half, they will flip the paper from which the figure was cut to create a symmetric image that uses positive and negative space to create the visual effect.

Management
1. Select colors of construction paper that make sense for the figures provided. Each figure requires two contrasting colors of paper. For the dragonfly and the butterfly, cut some of the construction paper in half horizontally. For the tree and the leaf, cut some of the construction paper in half vertically. Each student needs one full sheet of construction paper and one half sheet in a different color.
2. To make the templates, copy the figures and cut them out of file folders so that the straight edge of the figure aligns with the folded edge of the folder, as shown. To facilitate tracing, students can place their half-sheets of construction paper inside the file folders and tape them in place. Make multiple file folder templates for each figure.

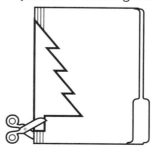

GETTING INTO GEOMETRY   123   © 2010 AIMS Education Foundation

3. Set the construction paper and file folder templates on a table where students can come and collect what they need.
4. Make an example of each figure so that students will have a model to look at while creating their own symmetry pictures.
5. It is expected that students have had several experiences with symmetry prior to doing this activity.

Procedure
1. Ask students how you can tell if something has symmetry. [If it has two halves that are mirror images of each other, it has symmetry.]
2. Challenge students to identify some things in the room that have symmetry. Have them justify their responses.
3. Tell students that they will be doing a symmetry art project. They will be making pictures of things in nature that have symmetry. Show them the examples that you created and have them identify each one.
4. Explain that students will be choosing one of these four figures to make—a tree, a leaf, a butterfly, or a dragonfly. To make a symmetry picture, one whole piece of construction paper and one half piece of construction paper in a different color are needed. Show students where the supplies are located and allow each student to choose a shape and select two colors of paper. (Students may need to share templates.)
5. Once each student has the necessary supplies, show students how to place their half sheets of construction paper inside the file folders so that the longer edge of the construction paper lines up with the inside of the crease in the file folder. Have them tape the file folders to the table and trace the figures.

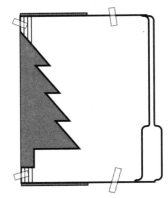

6. Distribute scissors and glue sticks and have students very carefully cut out their shapes along the lines they traced. Have them align the corners and edge of the border piece with the corners and edge of the full sheet of construction paper and glue it in place.

7. Using your examples as guides, have them glue down the figure they cut out so that it completes the other half of the symmetry picture.

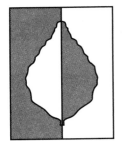 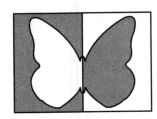

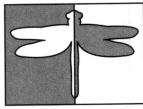

8. Discuss students' creations. Have them identify the lines of symmetry in their pictures and explain how they know the figures have symmetry.

Connecting Learning
1. What does it mean when something has symmetry? [It has two halves that are mirror images of each other.]
2. What are some things in our classroom that have symmetry? How do you know?
3. What did you have to do to make your symmetry pictures? [We cut out half of a picture.]
4. How do you know your picture has symmetry? [Both halves are mirror images of each other.]
5. Could you create other symmetry pictures using this technique? What would you have to do?

Extensions

1. Any shape or figure can be made into a symmetry picture by following the steps described in this activity. Allow students to cut out their own shapes from half-sheets of construction paper and see what they make when turned into symmetry pictures.
2. Make the symmetry pictures two-sided by cutting out the same figure from two half-sheets of paper and putting one on the front of the whole sheet and one on the back. Punch a hole in the top of the paper and hang them from string to make symmetry spinners.

* Reprinted with permission from *Principles and Standards for School Mathematics*, 2000 by the National Council of Teachers of Mathematics. All rights reserved.

The Art of Symmetry

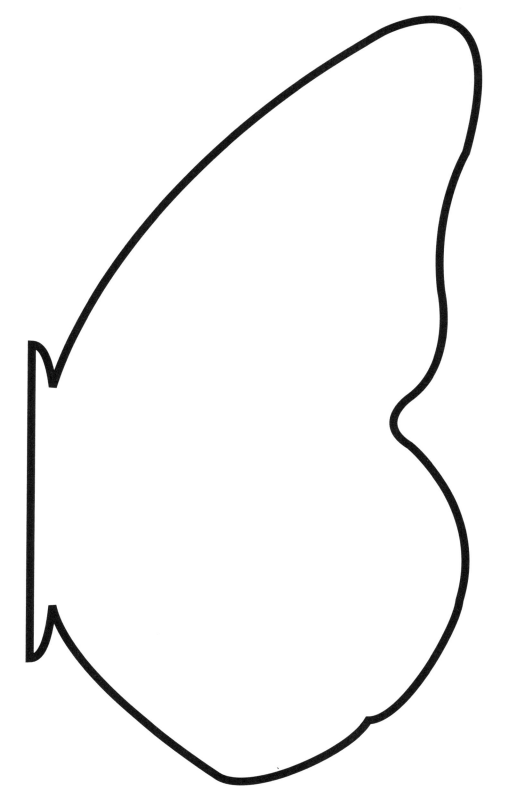

The Art of Symmetry

GETTING INTO GEOMETRY © 2010 AIMS Education Foundation

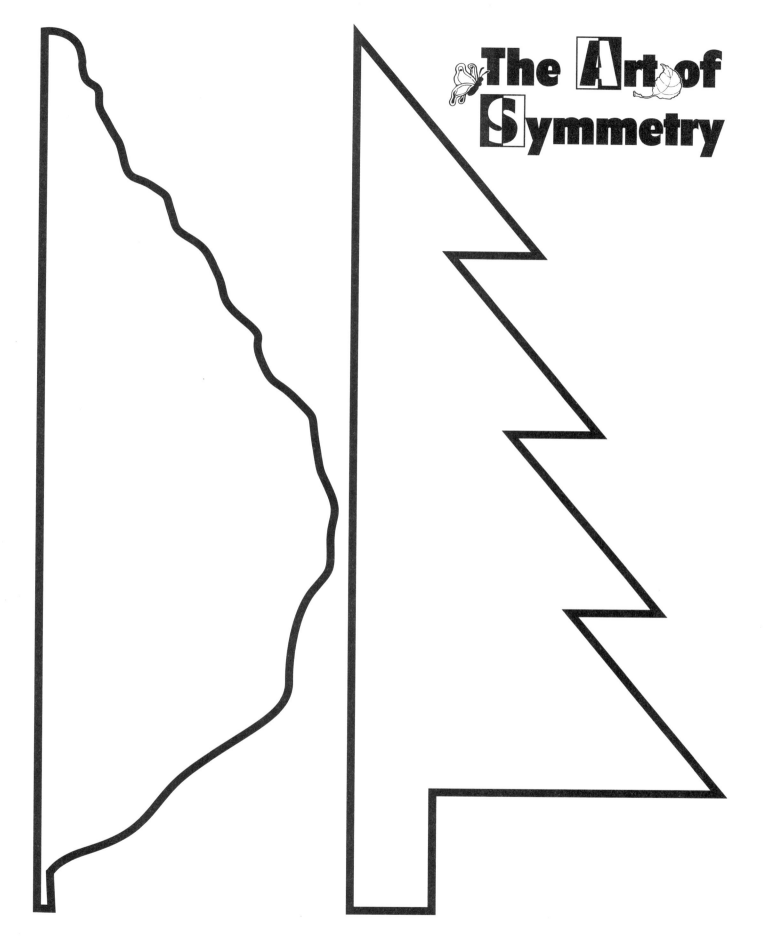

3-D Explorations

Topic
3-D solids

Key Question
How are geometric solids alike and different?

Learning Goals
Students will:
- sort three-dimensional objects by observable properties, and
- identify real-world examples of the 3-D solids.

Guiding Document
NCTM Standards 2000*
- *Recognize, name, build, draw, compare, and sort two- and three-dimensional shapes*
- *Describe attributes and parts of two- and three-dimensional shapes*
- *Sort and classify objects according to their attributes and organize data about the objects*

Math
Geometry
    3-D solids

Integrated Processes
Observing
Recording
Comparing and contrasting
Communicating

Materials
For the class:
    3-D solids (see *Management 1*)
    real-world geometric solids (see *Management 2*)
    chart paper

Background Information
   It is important for young learners to have multiple opportunities to explore three-dimensional geometric solids. These explorations should include experiences in which students are comparing models to real-world objects, real-world objects to pictures, and models to pictures. In this activity, students will compare real-world objects to models as they sort three-dimensional solids by observable properties.

Management
1. A manufactured set of six three-dimensional models, called Geo-solids, is available from AIMS (item number 4610). This set of includes a sphere, a cone, a cylinder, a cube, a rectangular solid, and a square pyramid.
2. Gather a variety of real-world examples of each of the three-dimensional solids you will be introducing with this activity. If desired, send a letter home with students requesting help gathering the items. Ideas include cereal, cracker, and tissue boxes; cylindrical chip cans; tennis ball cans; canned foods; balls in a variety of sizes; snow cone cups; party hats; and cake decorating tips. Place these objects around the room in full view of the students.

Procedure
Part One
1. Place a set of 3-D solids at the front of the classroom. Show one solid and tell the students its geometric name. Ask the students to gather real-world examples of this geometric solid from around the room. Go through all the solids following this same procedure.
2. Gather the students together in a circle, directing them to place the objects they found in the center of the circle.
3. Invite a student to observe the objects and sort them. Have him/her share the rule for the sort.
4. Place the objects in columns to create a concrete graph and discuss the data. For example, there may be three more objects that roll than objects that do not roll, etc.
5. Invite a second student to observe the objects and sort them by a different property. This time, don't have the student identify the rule he/she used for the sort; instead, ask the class to look at the groups of objects and try to determine the rule by looking at the properties of the solids in each group.
6. Direct a third student to sort according to shape. Invite the student to tell why he/she chose to place certain shapes together.
7. End with a discussion about the similarities and differences within the categories they have chosen. For example, in the rectangular solid set, some are short and fat, some are tall and thin, etc.

GETTING INTO GEOMETRY

Part Two
1. Gather the students together, giving each student one of the real-world objects to hold.
2. Using the following directions, play a listening game with the students. "If you are holding a shape like this (hold up a rectangular solid model), stand up. If you are holding a shape like this (hold up a sphere model), turn around. … like this (a cone model), stomp your feet." Continue this procedure, holding up an example of all the shapes being used in the game.
3. Have the students place the real-world solids behind them. Place several of the solid models in the center of the circle.
4. Ask for someone in the circle to find the shape with one flat face. …two flat faces. …six flat faces. Etc. Continue asking about shapes that roll, shapes that slide, etc.
5. Place all the rectangular solids that students found in the center of the group of students.
6. Ask students to generate a list of all the words they would use to name these objects (i.e., boxes, bricks, blocks, etc.). Record their responses on chart paper.
7. Use an analogy to introduce the vocabulary of geometric solids. Explain that each student has a name given to him/her by his/her parents (Taylor, Emily, Carson, Max, etc.). In the same way, these items all have names given to them by people (boxes, bricks, blocks, etc.). While names like Taylor, Emily, Carson, and Max are specific to certain kids, they can all be called children. It's the same way with these shapes. The special name that mathematicians have for all these shapes is *rectangular solids*. A cereal box is a rectangular solid. So are a brick, a block, and a box of crayons.
8. Print the word *rectangular solid* on the top of a piece of chart paper. Hold up a geometric model of a rectangular solid. Discuss with the students how this object is also a rectangular solid. Add this object to their collection.
9. Ask the students to make statements as to why all of these objects are called rectangular solids. What can we say that is true about all of them? [They all have flat faces. They all have eight corners. They all have edges. Their faces are different sizes. Their faces are all rectangles. Etc.] Record students' responses on the chart paper.
10. Direct the students to try to roll a few of these objects back and forth between partners. Have them try stacking these objects on top of each other.
11. Ask, "What did you find out about these rectangular solids? How are they the same? How are they different?" [They do not roll. They stack on top of each other.] Record on the chart.
12. Hold up a rectangular solid and ask the students to identify the solid you are holding. Turn the rectangular solid to the side. Ask, "Is it still a rectangular solid?" Turn it again and ask, "Is it still a rectangular solid?" Be sure students can articulate that no matter how it is turned, it always stays a rectangular solid.
13. Take a "rectangular solid" walk. Ask students to point out rectangular solids in the classroom and outdoors.
14. Repeat the process of identifying the characteristics of geometric solids with the cube, the cone, the sphere, and the cylinder. If desired, include the square pyramid.

Connecting Learning
Part One
1. What are some of the ways that we sorted the solids? How were these alike? How were they different?
2. How would you sort the solids? Why?
3. When we sorted by shape, what were some of the ways that all the solids in one group were alike? …different?

Part Two
1. What are the special names that mathematicians have for all of the solids we have been exploring? [rectangular solid, cube, cone, sphere, cylinder]
2. Which geometric solids have only one flat face? [cones] …two flat faces? [cylinders] …six flat faces? [cubes and rectangular solids] …no flat faces? [spheres]
3. What can we say that is true about all cubes? [They all have flat faces. They all have corners. They all have edges. All their faces are the same size. All their faces are squares. They do not roll. They stack on top of each other.]
4. What can we say that is true about all cones? [They all have a curved surface. They have a flat face. They all have a point.]
5. What did you find out about these cones? How are they the same? How are they different? [They can all roll. They can stack on top of each other if you hold one, but not by themselves. They can stack on their flat faces, but not on their curved surfaces.]
6. What can we say that is true about all spheres? [They are round. They have no corners. They have no edges.]
7. What did you find out about these spheres? How are they the same? How are they different? [They can all roll. They are all round. They do not stack on top of each other. Some are larger than others.]

8. What can we say that is true about all cylinders? [They are all round and flat on two faces. They have a curved surface.]
9. What did you find out about these cylinders? How are they the same? How are they different? [They can all roll. They can stack on top of each other on their flat faces, but not on their curved surfaces.]

Extension
Sing *The Shape-Up Song*.

Home Link
Ask the students to either bring in or draw pictures of objects they see at home that are the same shape as one of the solids they explored. Store these objects together in a container and leave them out for free exploration.

* Reprinted with permission from *Principles and Standards for School Mathematics*, 2000 by the National Council of Teachers of Mathematics. All rights reserved.

The Shape-Up Song
Additional Verses

Version #1 (Acting out)

A sphere is the shape that I'm being...
No matter how you look at me, I'm always round...

A cylinder's the shape that I'm being...
I'm round at both ends and I'm curved on the sides...

A cube is the shape that I'm being...
I've got corners, edges, and six flat faces...

(Guessing game)

This is the shape that I'm being...
(act out and give word clues)...

Version #2 (Choose a shape, sing appropriate verse)

A cone is the shape that I'm choosing,
A cone is the shape that I'm using.
An ice cream cone or a great party hat:
A cone is the shape that I'm choosing.

A sphere is the shape that I'm choosing...
Big ball, little ball, oranges, and beads...

A cylinder's the shape that I'm choosing...
Big cans, little cans, pipes, and tubes...

A cube is the shape that I'm choosing...
Blocks and boxes to keep things in...

Celebrating Solids

Topic
3-D solids

Key Question
How are geometric solids similar to the real-world objects at our party?

Learning Goal
Students will compare and contrast models of geometric solids to real-world objects such party hats, candles, and presents.

Guiding Documents
Project 2061 Benchmarks
- *Numbers and shapes can be used to tell about things.*
- *Use numerical data in describing and comparing objects and events.*

NCTM Standards 2000*
- *Recognize, name, build, draw, compare, and sort two- and three-dimensional shapes*
- *Describe attributes and parts of two- and three-dimensional shapes*
- *Sort and classify objects according to their attributes and organize data about the objects*
- *Represent data using concrete objects, pictures, and graphs*

Math
Geometry
    3-D solids
Sorting
Graphing

Integrated Processes
Observing
Classifying
Collecting and recording data
Comparing and contrasting
Interpreting data

Materials
Party items (see *Management 2*)
Birthday candles (see *Management 3*)
Geo-solids (see *Management 1*)
Three-dimensional shaped snacks (see *Management 6*)
Student pages
Party station cards

Background Information
Children often learn to recognize and name a wide variety of two- and three-dimensional shapes and solids at a very young age. This activity will provide an opportunity for them to extend their understanding of 3-D solids by comparing and contrasting them with real-world objects. A birthday scenario will provide playful real-world objects for students to observe. As students compare the chosen items to models, they will begin to recognize similarities in other real-world objects as well.

Management
1. Geo-solids are available from AIMS (item number 4610).
2. Gather an assortment of party items in the shape of the Geo-solids—cone, cube, square pyramid, sphere, rectangular solid, and cylinder. For example: candles, party hats, round balloons, party horns, etc.
3. In addition to regular birthday candles, purchase novelty birthday candles in several shapes, such as spheres (soccer balls, basketballs), cones (construction pylons), or cubes. Place a variety of candles in a box, and wrap the box like a birthday present. Make one box (present) for each group.
4. Groups of two or three students work well for this activity.
5. Prior to teaching this lesson, decorate the classroom with the various party items. Place a balloon, present, party hat, etc., at each table.
6. You may suggest to parents of children with summer birthdays that your class will be having a "un-birthday" party that will focus on three-dimensional geometry. Suggest that they can send party treats as long as they are shaped like geometric solids. Suggestions include: spheres—cheese puff balls, malted milk balls; cones—funnel shaped corn snacks; cylinders—mini-marshmallows, carrots, pretzel sticks; cubes—caramels; rectangular solids—snack-size candy bars, fruit flavored chews.
7. Prepare five snack stations for *Part Two*. Place the station cards and the corresponding snacks at stations through which students can rotate.

GETTING INTO GEOMETRY   © 2010 AIMS Education Foundation

Procedure
Part One
1. Tell the class that they will be participating in a "un-birthday" celebration. Explain that the celebration will focus on the three-dimensional solids that they have been studying.
2. Hold up a geometric solid such as a cube. Ask the students to find objects in the collection of birthday items that resemble the shape of this object. Discuss how they are similar and how they are different. Record student responses on the board under the title *cube*. Show students a model of a sphere. Ask them if they can find anything at their table that looks like it. [balloon]
3. Use the wrapped present to play a game of *Twenty Questions*. Challenge the students to try to figure out what might be inside. Each question needs to be answered with a yes or no. Teach students to ask questions that will lead them closer to discovering the contents. For example: Is it alive? [No.] Do we eat it? [No.] Do we wear it? [No.] Do we see them at a birthday party? [Yes.] Does it go on a cake? [Yes.] Is it candles? [Yes.] These questions lead the students to the answer. Questions that are too specific to narrow the choices would be: Is it red? Is it a cake? Is it a dog?
4. Once the students have guessed the contents, allow them to open the presents. Discuss the shapes they see. Give each group the sorting and graphing pages. Have students find one rule by which to sort the candles. Allow each group to share their rules, then have them sort again using rules that have not yet been suggested. Encourage them to use rules such as cubes—not cubes, cylinders with stripes—cylinders without stripes. Discuss their results.
5. Ask the students to sort their candles by shape one more time. Have them graph the information on the second student page.
6. Ask the students to compare their graphs with others. Discuss the results.

Part Two
1. Review the names of the Geo-solids by holding them up one at a time and asking the students to identify them and real-world examples of them.
2. Explain to students that they are now going to follow a set of directions and collect a shape snack.
3. Allow students to rotate through the snack stations.
4. When students have gathered their snacks, play "Everybody Show" by asking students to show you examples of each shape. For example, "Everybody show me a cube-shaped snack. Everybody show me a cone-shaped snack." Etc.
5. End with a discussion about other party objects that could have represented the shapes at today's party while they enjoy their snack.

Connecting Learning
1. What things did we do in this celebration that used shapes? What did you learn?
2. What rules did you use to sort the candles?
3. How did your rules compare to what other groups used?
4. Describe your graph. What does it tell you about your candles?
5. How does your group's graph compare to other groups' graphs? What does this tell you about all of the candles in the class?
6. What real-world party items were shaped like a sphere? ...cube? ...cone?
7. What shape is a cheese ball? ...balloon? ...present?
8. Where else can we find objects that are shaped like a sphere? ...cube? ...cone?

* Reprinted with permission from *Principles and Standards for School Mathematics*, 2000 by the National Council of Teachers of Mathematics. All rights reserved.

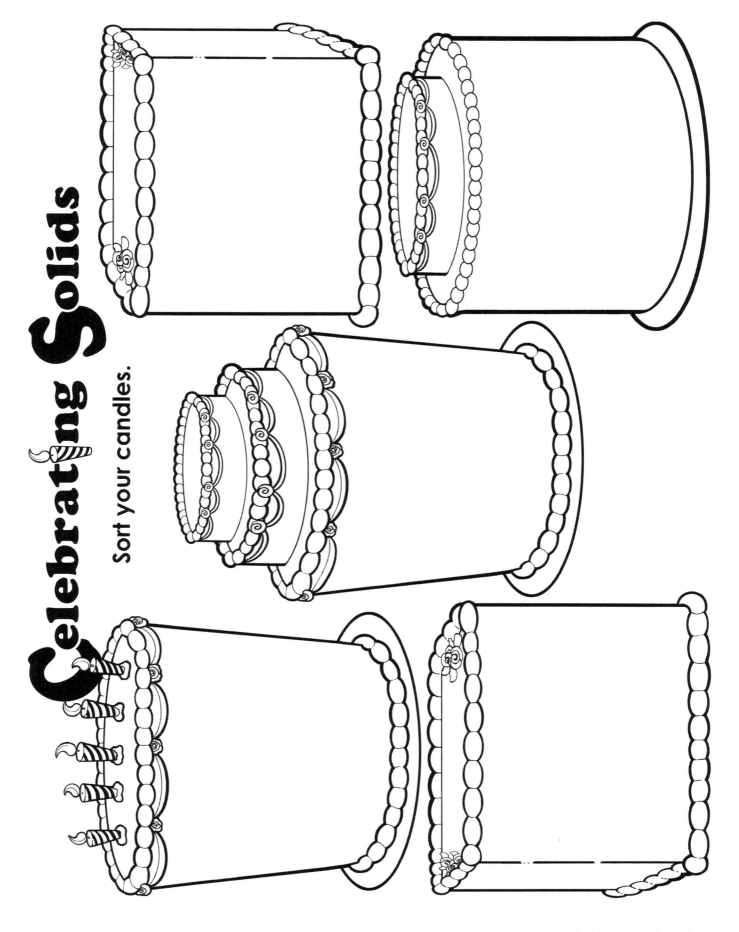

Solids on the Slide

Topic
3-D solids

Key Question
How does the shape of an object affect how it moves?

Learning Goal
Students will observe the effects of shape on how things move.

Guiding Documents
Project 2061 Benchmarks
- *Things move in many different ways, such as straight, zigzag, round and round, back and forth, and fast and slow.*
- *Describe and compare things in terms of number, shape, texture, size, weight, color, and motion.*

NCTM Standards 2000*
- *Recognize, name, build, draw, compare, and sort two- and three-dimensional shapes*
- *Describe attributes and parts of two- and three-dimensional shapes*

Math
Geometry
    3-D solids

Integrated Processes
Observing
Collecting and recording data
Comparing and contrasting
Communicating
Applying

Materials
For each group of students:
    collection of real-world objects (see *Management 1*)
    set of 3-D solids (see *Management 2*)
    ramp (see *Management 4*)
    chart paper

For each student:
    Solids on the Slide journal

Background Information
This investigation is designed to heighten an awareness in young children of how geometric solids move. Students need to observe, describe, and discuss moving solids including soccer balls, soup cans, and Geo-solids. They will explore questions such as: Does it move in a straight line? How many different ways does it move? How far does it move with one push? Does it move fast or slow?

Management
1. Ask the students to bring in a collection of various real-world items in the shape of spheres, cones, cubes, cylinders, and rectangular solids that are not cubes. Some appropriate items would be: balls, snow cone cups, tissue boxes, cereal boxes, tennis ball containers, etc.
2. A set of six three-dimensional models, called Geo-solids, is available from AIMS (item number 4610). This set of includes a sphere, a cone, a cylinder, a cube, a rectangular solid, and a square pyramid.
3. Copy the included pages to make one *Solids on the Slide* journal for each student. Cut the pages along the solid lines. Extend the left side of the book to the edge of the paper to allow room for stapling. Place the pages in order starting with the smallest tab on the first page and ending with the full pages at the back.

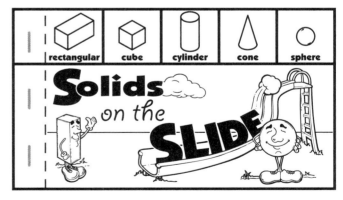

4. Have materials available for students to use as ramps. For example, a piece of cardboard or wood that is at least 150 centimeters in length and 50 centimeters wide. One end of the ramp can be placed on a chair. A table with two legs raised can also serve as a ramp. Discuss appropriate safty rules.

GETTING INTO GEOMETRY   143   © 2010 AIMS Education Foundation

Procedure
1. Divide the class into small groups. Give each group of students several real-world examples of cylinders, spheres, cones, cubes, and rectangular solids.
2. Give the students time to explore how the objects move on the floor. [fast, slow, straight, in a circle, slide, roll, etc.]
3. Gather the students together and discuss what they found out about their collections of objects.
4. Display the Geo-solids for everyone to see. Point to the sphere and ask the class which of their objects is most like the sphere. Repeat with each of the other Geo-solids.
5. Draw students' attention to the square pyramid. Write the name of the shape onto a piece of chart paper. Ask the students how they think the square pyramid would move by itself if it were placed on a hill or inclined plane. …on a flat surface. Question them about whether it would roll, slide, or flop, and whether it would move in a straight line or curve as it moves ahead.
6. When students have given their predictions about how the square pyramid would move, place it on a flat surface to see if it will move on its own. Then, apply a small amount of force and record on the chart paper how the shape moved. [slid, flopped, etc.] Place the shape on an inclined plane and repeat the process.
7. Distribute a *Solids on the Slide* journal to each student.
8. Have your students put their real-world objects in a designated place. Give each group of students a set of Geo-solids and explain that they will explore how the other solids move on both a flat surface and on an inclined plane. Ask students to use pictures and words to describe the movements. (It might be necessary to provide a word bank for students.)
9. After students have had sufficient time to explore, gather the class back together and discuss what they discovered. Spend some time physically sorting the solids as you discuss those that slid, those that rolled, etc.

Connecting Learning
1. Which solid moved fastest? …slowest? …farthest?
2. Which solids traveled in a straight line? …a curved line? How are they alike? How are they different?
3. Describe all the shapes that rolled. How do they compare to the shapes that did not roll? [All the shapes that rolled—the sphere, cone, and cylinder—have a curved surface.]
4. Did how you placed a solid on the surface affect how it moved? Explain. [It depends on the solid. No matter how you place a sphere, it should roll. If you place a cone or cylinder on its base, it should slide. If you place a cone or cylinder on its curved surface, it should roll.]
5. How did putting the solids on a ramp change how they moved?
6. Which shapes moved down the ramp without a push? Which shapes needed a push to get them to move down a ramp? Do you think a different ramp would give you different results? Explain. [Perhaps. A steeper ramp might cause some solids to move on their own that would not move on a ramp with a shallow incline, and vice versa.]
7. What shape do you think would be best to make something that rolls easily? Explain.
8. What shape would be best to make something that would stay in one place if it were on a ramp? Explain.

Extension
Ask the students to observe things at home that move. Tell them to report to the class the shapes of these objects and to describe what they notice about how they move. Do they need help to move? Do they move by themselves? Do they move fast or slow? Do they move in a straight line or a curved line?

* Reprinted with permission from *Principles and Standards for School Mathematics*, 2000 by the National Council of Teachers of Mathematics. All rights reserved.

rectangular solid

I discovered that a rectangular solid…

Make a Match

Topic
3-D solids

Key Question
Without using your eyes, how can you tell if two solids are alike?

Learning Goal
Students will:
* identify cylinders, spheres, and cubes by name;
* describe characteristics of these solids; and
* find a matching pair of solids using only the sense of touch.

Guiding Document
*NCTM Standards 2000**
* *Recognize, name, build, draw, compare, and sort two- and three-dimensional shapes*
* *Describe attributes and parts of two- and three-dimensional shapes*

Math
Geometry
   3-D solids

Integrated Process
Observing
Comparing and contrasting
Communicating

Materials
For each group of students:
   attribute beads in bags (see *Management 3*)
   mystery box (see *Management 1*)

Background Information
   Young children need multiple opportunities to use the vocabulary of geometry in order for them to internalize the terms. This activity provides a playful opportunity for students to explore the properties of a select group of three-dimensional solids at a tactile level. At this level, students should be able to identify the solid and describe some of its properties. For example, a cone has one flat face and one point; a cube has six flat faces that are equal in size; etc.

Management
1. Prior to teaching this lesson, prepare several mystery boxes, one for every three or four students. To prepare a box, remove the lid from a cardboard shoebox and cut out a hand-hole on each short side of the shoebox. If desired, cover with contact paper to add a decorative touch.

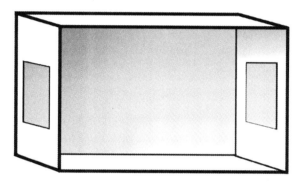

2. It is suggested that this activity follow experiences where the names and attributes of the solids have been taught, such as the activity *3-D Explorations*.
3. Prepare bags of attribute beads for each group of students. The bags should include at least two of each shape—cylinders, cubes, and spheres, Attribute beads and bags (item number 4122) are available from AIMS.

Procedure
1. Gather the class into a circle. Place one of each of the attribute beads in the center of the circle Invite the students to name each shape as you point to them.
2. Ask the class if they think that they could identify the shapes without being able to see them.
3. Tell them that they will put their hands behind their backs and close their eyes. You will then put an attribute bead in their hands and ask them to indentify the shape.
4. Invite a student to try. When the student says, "It is a ____.", ask why he or she didn't think it was (some other shape). This gives the student the opportunity to describe the characteristics of the specific shape.

GETTING INTO GEOMETRY © 2010 AIMS Education Foundation

5. Choose another student to try, but this time put a different shape in each hand. Ask the student if the shapes are the same. Do they match?
6. Divide the class into small groups. Give each group a set of attribute beads and a mystery box. Tell the students that they will select one bead of each shape to put under the mystery box. Then they will select at least one bead that will make a matching set of shapes to also put under the box.
7. Instruct them that one student in each group will close his or her eyes. Then another group member will place the beads on the table and turn the box so that it covers the beads.
8. Direct the group members to help the students with their eyes closed to place their hands in the holes on the sides of the boxes.
9. Challenge the students to feel the beads under the box and find two matching shapes. Tell them to place one of the two shapes in each hand and remove their hands from the box to see if they have a matched set.
10. Repeat the procedure with each group member having several opportunities to make matches, each time changing the shapes that match under the box.
11. End with a discussion about how students decided whether the shapes were a match when they could not see them.

Connecting Learning
1. How did you find a matching pair of solids without being able to see them?
2. Which solids were the easiest to match? ...hardest? Why?
3. Do you think it is easier to find matching solids using your eyes or your hands? Why?
4. Show me a cube. ...cylinder. Etc.
5. Could you find two red cubes under the mystery box? Why or why not?

* Reprinted with permission from *Principles and Standards for School Mathematics*, 2000 by the National Council of Teachers of Mathematics. All rights reserved.

Behind the Curtain

Topic
3-D solids

Key Question
What 3-D solids were you shown?

Learning Goal
Students will briefly look at a selection of 3-D solids and try to recall what they saw.

Guiding Document
NCTM Standards 2000*
- *Recognize, name, build, draw, compare, and sort two- and three-dimensional shapes*
- *Create mental images of geometric shapes using spatial memory and spatial visualization*

Math
Geometry
    3-D solids
    spatial memory

Integrated Processes
Observing
Comparing and contrasting
Communicating

Materials
3-D solids (see *Management 1*)
Large towel (see *Management 2*)

Background Information
    How often have you asked questions such as: Where did I leave my keys? Do you know where I put the scissors? In order to help you find a lost item, others must know what the object is (or what it looks like) and recall where they saw it. When they can answer the question that ends your search, they are using spatial memory to recall the location and the object.
    This activity is designed to help students strengthen visual memory skills using familiar 3-D solids. Not only will they rehearse the naming of the objects, they will develop strategies for helping them recall a missing object from a set previously observed.

Management
1. Geo-solids are available from AIMS (item number 4610). Have two or three sets available so you can display two or more of the same solids. For example: two cubes, two cones, and a sphere.
2. A large towel will serve as a curtain that students will hold up in front of the solids in order to hide them.

Procedure
1. Show the students the set of 3-D solids. Ask students to name the solids and compare and contrast their attributes.
2. Inform students that you are going to place some of the solids on the table and then secretly remove one. It will be their challenge to tell you which solid was removed from the group.
3. Set out three or four solids in an area where all students can see. Invite two students to hold up the large towel to hide the solids.
4. Remove one of the solids and put it where students can't see it.
5. Tell the students holding the towel to lower it to allow the class to see the remaining solids.
6. Ask if anyone thinks they know which solid is missing.
7. Continue the procedure, changing the numbers and positions of the solids. Have students take turns holding the towel so that all can participate in determining the missing solids.
8. Urge the students to share the various strategies they used to determine the missing solids.

Connecting Learning
1. How did you remember what solids you saw?
2. Were there any solids that were easier to remember than others? Explain.
3. Do you think you got better at remembering the solids as we went along? Explain.
4. If you close your eyes, can you imagine the cube? What does it look like? How is it different from the cylinder?
5. What other solids can you imagine when you close your eyes? Describe them.

Extension
Follow the same procedures but replace the geometric solids with real-world examples, such as a soup can for a cylinder, a ball for a sphere, etc.

* Reprinted with permission from *Principles and Standards for School Mathematics*, 2000 by the National Council of Teachers of Mathematics. All rights reserved.

All About Town

Topic
3-D solids

Key Questions
1. What kind of town can we build using geometric solids?
2. Can we build another group's town using only the "footprints" of their town?

Learning Goals
Students will:
- use three-dimensional solids to build a town, and
- challenge another group to recreate their town using the footprints of their creation.

Guiding Document
NCTM Standard 2000*
- *Investigate and predict the results of putting together and taking apart two- and three-dimensional shapes*

Math
Geometry
   3-D solids

Integrated Processes
Observing
Comparing and contrasting
Communicating

Materials
For each group:
   geometric solids (see *Management 3*)
   chart paper
   marker

For the class:
   digital camera
   printer

Background Information
Children need time to manipulate and explore the geometric solids they find in their own world. This exploration allows them to both become aware of the characteristics of each solid as well as the attributes of the plane figures that make up each solid.

Management
1. Students should work in groups of four or fewer when constructing their own towns.
2. Desks, tables, and chairs may need to be moved to allow each group adequate floor space for building their towns.
3. Use real-world geometric solids previously collected or send a letter home asking families to send in boxes, cans, and other containers.
4. Build a town and trace its footprint prior to doing this activity. This will be used in the introduction of this activity. Objects should not be stacked.
5. A digital camera is used to take a picture of each town before disassembling it. You will also need to be able to print the pictures you took to give groups as a reference.
6. In the part of the activity where students recreate each other's towns, it is very possible to find many different ways to construct the town with the same footprint. This should be recognized as acceptable and not incorrect. Students should be encouraged to feel good about their solutions and curious as to how the other groups built the town.

Procedure
1. Gather students to an open space area. It is preferable for students to be seated in a circle so all can see. Show them the footprint of your town and the various geometric solids you used to build your town. Ask them how they think you built your town using these solids.
2. Allow several students to try to construct a town like yours.
3. Discuss the possible ways the town could be built and then show them your actual construction.
4. Demonstrate to the students how you created your town and then made the footprint of it by tracing around the bases of the solids. For spheres, students will need to draw circles.
5. Distribute the materials and challenge students to work with their groups to create a town.
6. Allow time for students to explore the building of the town and then encourage them to begin making its footprint. During this stage they may wish to label their buildings (e.g., water tower, park, police station).

GETTING INTO GEOMETRY   © 2010 AIMS Education Foundation

7. After groups have constructed their towns, take a picture of the construction using a digital camera. Then, have them remove the solids from the footprints and stack them randomly to the side of their tracings.
8. Have each group trade places with another group and try to build their town using that group's footprints and solids.
9. Print the pictures of the groups' towns and set them aside for later.
10. As groups are working to reconstruct the towns, they may want to hire a consultant (someone who actually built the town) to consult with them and give them hints as to how the town was made. One student from the construction crew can be asked to consult for a few minutes.
11. Once groups have completed building, have them return to their sites and give them the pictures of their original constructions. Have them compare the recreated towns with their originals. Give groups time to compare and discuss the rebuilding process with each other.
12. Bring the class together and discuss the process.

Connecting Learning
1. How were the towns similar? …different?
2. What were the solids that were easiest to use in your construction? Why? …hardest? Why?
3. What are some other solids you wish you would have been able to use when building?
4. What did you learn about the footprints of different solids?
5. What did you learn from doing the building?
6. What did you like about building the town? …dislike?

Extensions
1. Once the reconstruction challenge is complete, students may want to paint their towns, create roads and trees, and then put all towns together to create a large city.
2. Take students on a walk around school or the neighborhood to find solids in the buildings and structures they see on the walk.

*   Reprinted with permission from *Principles and Standards for School Mathematics*, 2000 by the National Council of Teachers of Mathematics. All rights reserved.

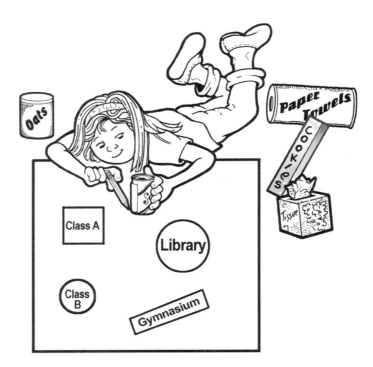

Same Shapes

Topic
3-D solids

Key Question
What would the picture of a cube look like?

Learning Goal
Students will name and match three-dimensional solids to their pictures.

Guiding Document
*NCTM Standards 2000**
- *Recognize, name, build, draw, compare, and sort two- and three-dimensional shapes*
- *Describe attributes and parts of two- and three-dimensional shapes*

Math
Geometry
   3-D solids

Integrated Processes
Observing
Comparing and contrasting
Communicating

Materials
For each student:
   shape cards

For the class:
   Geo-solids (see *Management 1*)
   Solid Shape Mats

Background Information
It is important for students to make connections between three-dimensional models and two-dimensional picture representations of those models. This activity allows for students to see and touch the models while looking for the picture that best represents what is in front of them.

Management
1. Geo-solids can be purchased from AIMS (item number 4610).
2. Copy one set of the six shape cards for each student—cube, cone, cylinder, rectangular prism, sphere, and square pyramid.
3. Copy one set of the *Solid Shape Mats* to be placed in a center at the conclusion of this lesson.

Procedure
1. Gather the class into a circle. Place the set of the Geo-solids in the center of the circle. Review the names of the solids.
2. Give each student a set of shape cards. Invite a student to come to pick up the sphere. Tell the student to make sure that each student can clearly see what he/she has chosen. Ask the rest of the class to find the picture in their set of cards that looks most like the sphere. Have them place the card on the floor in front of them. Question the students about the characteristics of the sphere, and how they knew which picture matched it.
3. Repeat this procedure with the other solids.
4. When all solids have been addressed, describe a solid and ask students to hold up the shape card of the matching solid. For example: Which shape has only square sides?
5. Have students turn their cards face down. Select one student to turn one card over. Tell that student to call out the name of the solid that is indicated on the card. Have the student who is to the right of the card turner select the represented solid.
6. Continue around the circle with the next student to the right turning over the card.
7. End by showing students how to use the *Solid Shape Mats* and explain that the Geo-solids and the mats will be placed in a center for extra practice.

Connecting Learning
1. What does the picture of a cube look like? ...sphere? ...rectangular solid? ...cone? ...cylinder?
2. Which shape was easiest to match to its picture? Why do you think that is?
3. Which shape has two circular ends? ...all six sides the same?
4. What two solids are most alike? [the cube and the rectangular solid] How are they alike? [They each have six flat faces. Each end is a square.] How are they different? [The rectangular solid has some faces that are not squares. The faces of the cube are all squares.]
5. Where have you seen pictures of Geo-solids?

* Reprinted with permission from *Principles and Standards for School Mathematics,* 2000 by the National Council of Teachers of Mathematics. All rights reserved.

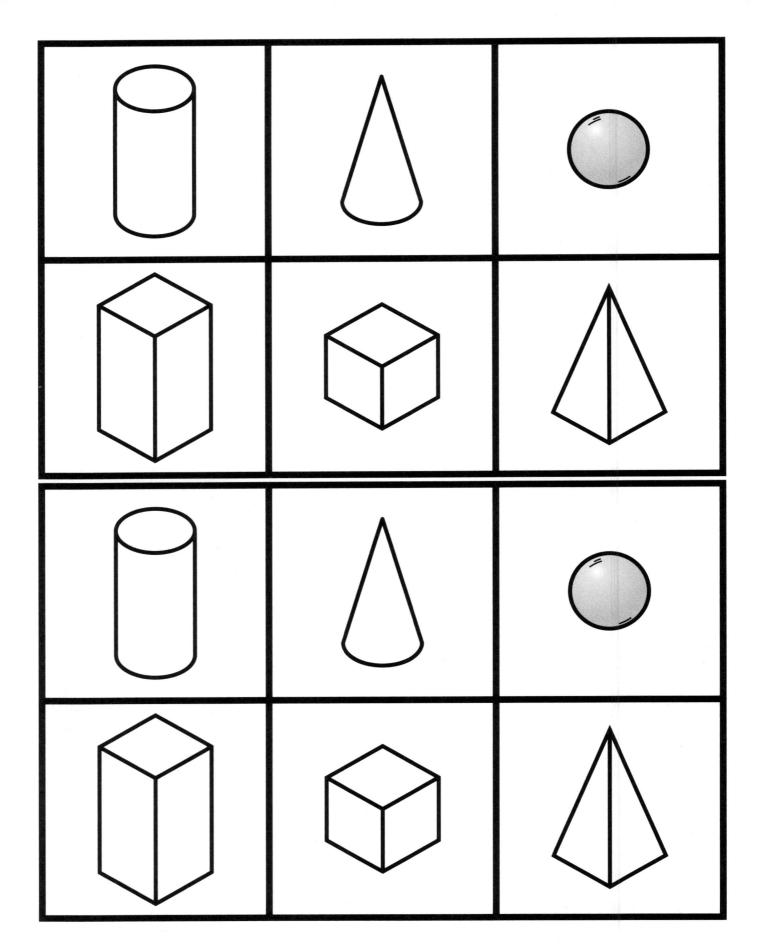

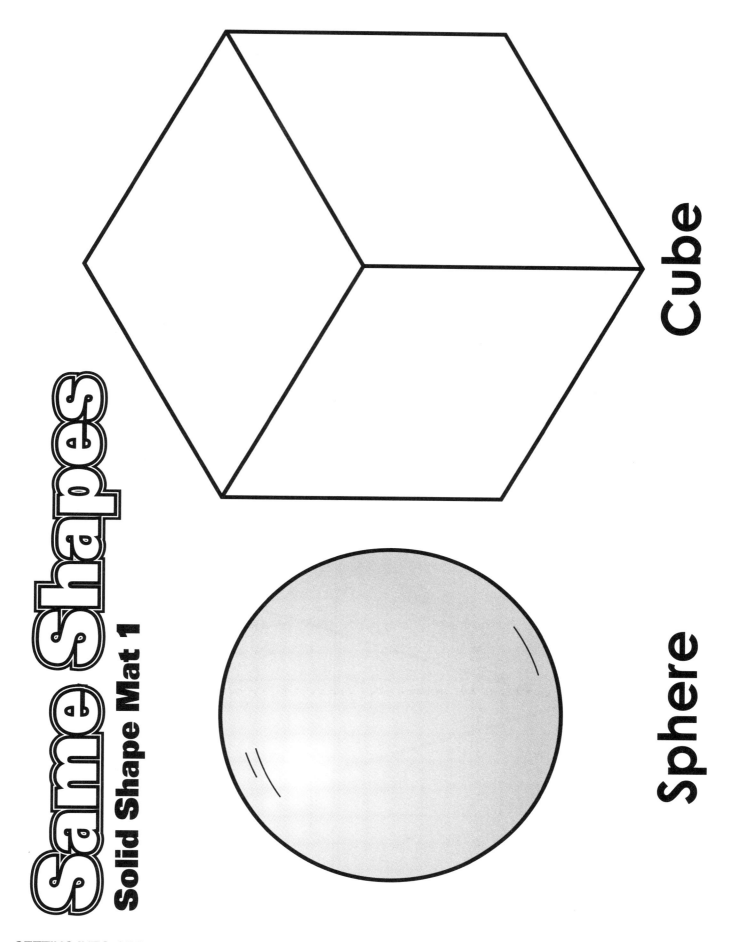

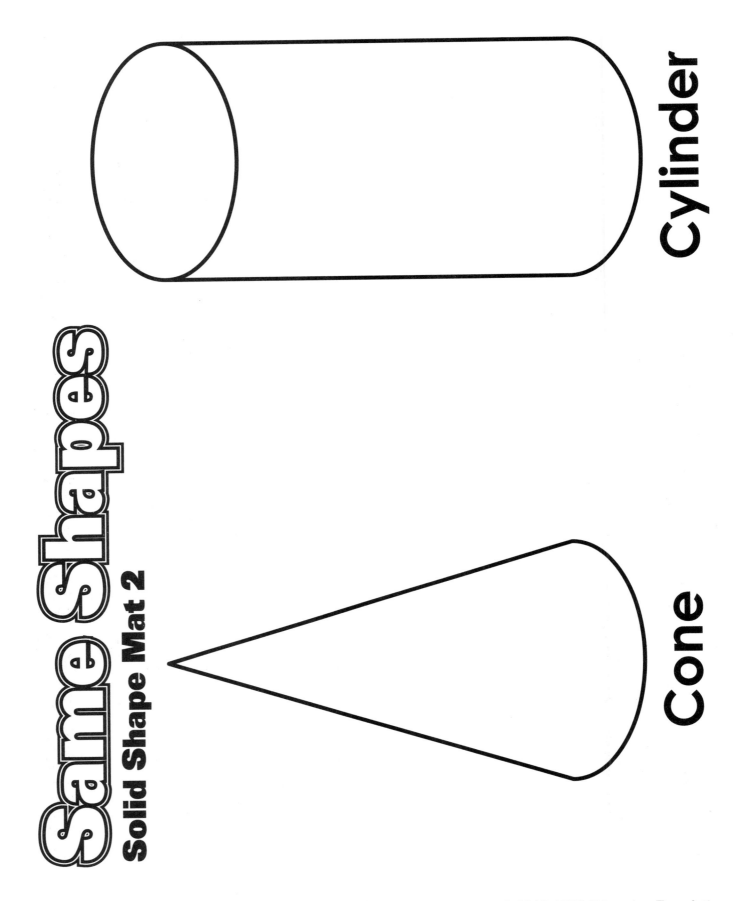

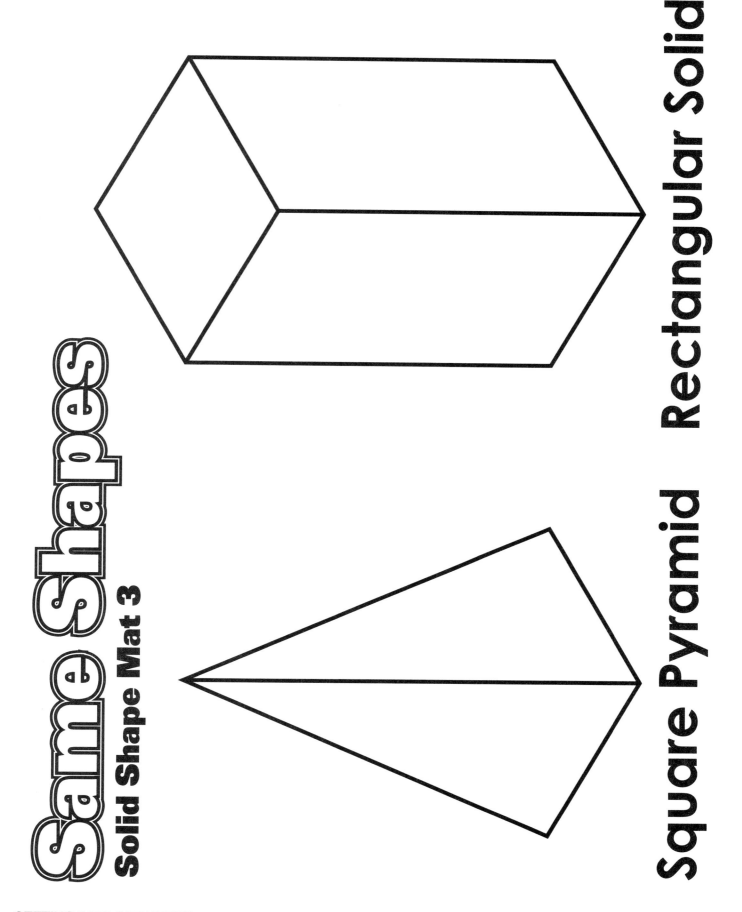

Rootin' Tootin' Relay

Topic
3-D solids

Challenge
Students will match a shape card to the three-dimensional object/solid it represents.

Learning Goal
Students will play a game during which they will match pictures of 3-D solids to Geo-solid models and/or real-world examples of the 3-D solids.

Guiding Document
NCTM Standard 2000*
- *Recognize, name, build, draw, compare, and sort two- and three-dimensional shapes*

Math
Geometry
    3-D solids

Integrated Processes
Observing
Comparing and contrasting
Identifying
Communicating

Materials
Geo-solids (see *Management 2*)
Various real-world examples of solids
    (see *Management 2*)
Shape cards, one per student (see *Management 3*)

Background Information
    Young children need repeated experiences in order to develop a firm foundation of geometry. Using geometric solids from the real world, as well as models and pictures of geometric solids, provide opportunities for students to make connections among the various representations of the solids.
    This activity is designed to engage students in a physical game while they are making connections between 2-D pictures and their 3-D equivalences. They will play a relay-type game in which they run to a given area, flip over a card, select the matching 3-D solid, and return to their team with both picture and solid.

Management
1. It is assumed that students have had repeated experiences identifying solids before playing the game.
2. A large and varied collection of geometric solids should be gathered prior to the activity. Both the AIMS Geo-solids (item numbers 4610 or 4612) and real-world examples of those solids should be used. Try to gather a variety of sizes of the same shape. You will need at least one object for every person in the class. Objects used are cubes, cylinders, spheres, cones, square pyramids, and rectangular prisms.
3. Prior to teaching this lesson, make sets of the shape cards so that each player on the team will have a card to turn over. Gather the corresponding numbers of 3-D solids, plus a few more so that the students near the end of the race still have objects from which they must select a matching solid.
4. Choose a table on which to put the solids and the shape cards. Count out the number of shape cards needed for each of two teams. Place the cards face down on opposite sides of the table—one set for one team, the other set for the other team. The solids should be between the two card sets.

Procedure
1. Review the names of the solids students will be using in this relay and have students share the characteristics of these solids. [Spheres are round, cubes have square faces, cones have a pointed end and a base that's a circle, etc.]
2. Divide the students into two teams.
3. Show the teams the collections of geometric solids and the areas where they will match the pictures to those solids.
4. Have one student model running to the front display, flipping over a card, gathering the corresponding solid, and running back with the card and the solid to be checked by you. When okayed, tell them that the relay runner will tag the next person on his/her team. If the solid does not match the picture, the relay runner must return the solid to the table and find the one that actually matches.
5. Explain that when each member of the team has run, correctly matched an object, and returned to the line, that team is finished. The first team to finish is the winner.

GETTING INTO GEOMETRY

6. Make sure that teams understand from which side of the table they are to select their cards.
7. After playing the game several times, take the objects and organize them into a real graph. Discuss how many they have of each solid, which they have more of, etc.

Connecting Learning
1. How did you decide which solid matched your picture?
2. Of which solid did we have the most? Why might this be the case? [There are more examples of that type of solid in the real world.]
3. Did each group start with the same number of picture cards? Why? [Yes, the game would not be fair if one group had to match more solids.] Did both groups have the exact same set of pictures? How do you know this? [Answers might include something like: No, one group has more cubes than the other so they would have had more cube pictures.]
4. Are there any shapes we have studied that had very few examples of in the collection? Why or why not? [Square pyramid. We didn't have a lot of real-world examples of that shape.]

* Reprinted with permission from *Principles and Standards for School Mathematics*, 2000 by the National Council of Teachers of Mathematics. All rights reserved.

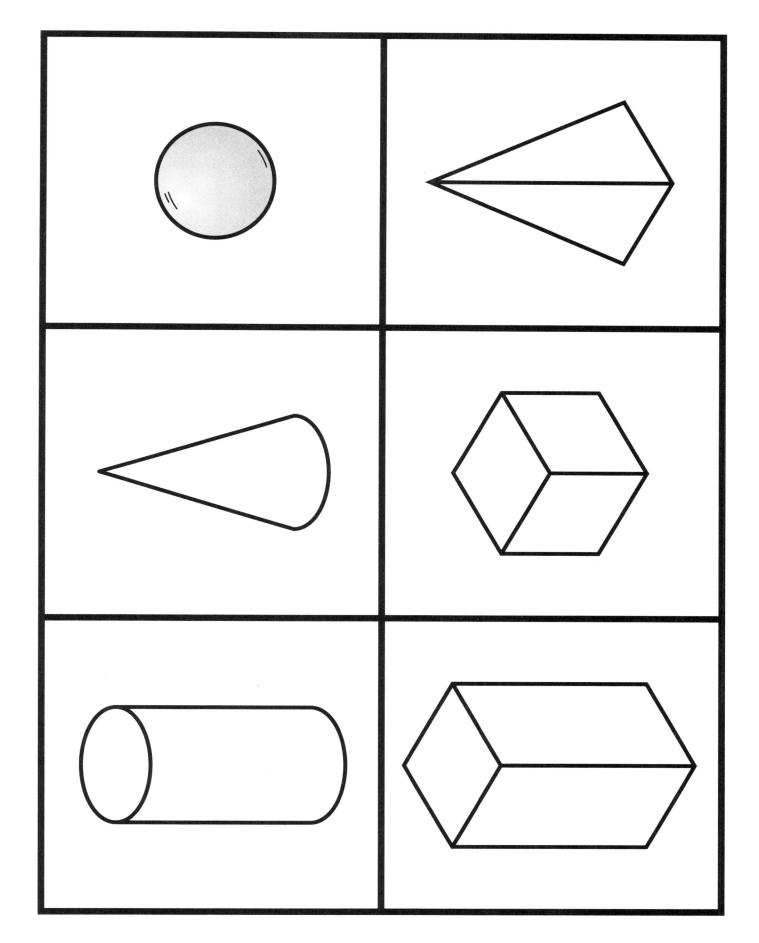

Solid Combinations

Topic
3-D solids

Key Question
How can we combine a set of three-dimensional solids to make new shapes?

Learning Goal
Students will attempt to build pictured combinations of three-dimensional solids to determine if they are possible.

Guiding Document
NCTM Standards 2000*
- *Recognize, name, build, draw, compare, and sort two- and three-dimensional shapes*
- *Investigate and predict the results of putting together and taking apart two- and three-dimensional shapes*

Math
Geometry
   3-D solids

Integrated Processes
Observing
Identifying
Predicting

Materials
Geo-solids (see *Management 1*)
Student pages

Background Information
    An important geometric idea for young children to explore is that solids can be combined or subdivided to make other solids. For example, two cubes combine to form a rectangular prism. By putting shapes together and taking them apart, children deepen their understanding of the attributes of the solids and how they are related. Do they stack? Which solid would fit exactly on top of the cylinder? Etc.
    In this activity, students are asked to look at two-dimensional drawings of three-dimensional solids and decide if specific combinations can be made. They will then use three-dimensional models to check their thinking. Moving back and forth between 3-D objects and their 2-D representations will help students recognize the shapes being represented and give them a better understanding of the characteristics of common 3-D solids.

Management
1. Each pair of students will need a set of Geo-solids. Geo-solids are available from AIMS (item number 4610). The set includes a cube, a rectangular prism, a square-based pyramid, a cylinder, a cone, and a sphere. These are the solids pictured on the student pages.
2. This activity can be done with the entire class and then placed at a learning center for additional practice.

Procedure
1. Display a cylinder. Ask students to name the solid. Have them identify other similarly shaped objects in the classroom. Question the students about whether the cylinder can easily stack or combine with additional cylinders or other solids.
2. Give each pair of students the first student page and a set of Geo-solids. Invite each pair to select the two solids pictured at the top of the page. Ask students if they think they could combine the two solids as pictured in the diagrams. Have students predict whether the combinations are possible and then test their thinking using the solids.
3. Give each pair the additional student pages and allow time for them to predict and test the combinations pictured.
4. End with a discussion about how they decided if a combination was possible.

Connecting Learning
1. What are some things that you observed about the solids?
2. How are the cube and rectangular solid the same? How are they different?
3. How did you decide if a combination was possible?
4. Were there any combinations that you thought would work until you tried them?
5. Which solids stack well? Which do not?
6. If you wanted to build a very high tower, what solids would be best to use? Why?
7. Do you think you could use all of the solids to make a single tower? Explain.

* Reprinted with permission from *Principles and Standards for School Mathematics*, 2000 by the National Council of Teachers of Mathematics. All rights reserved.

GETTING INTO GEOMETRY

Solid Combinations

What can you do with these?

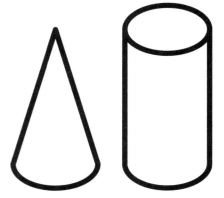

Can you do this?   Can you do this?

           Yes or No   Yes or No

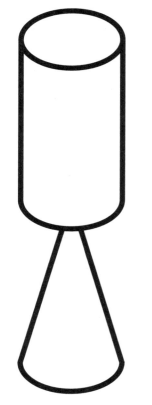

What else can you do?

GETTING INTO GEOMETRY

Solid Combinations

What can you do with these?

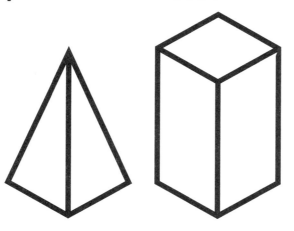

Can you do this?

Yes or No

Can you do this?

Yes or No

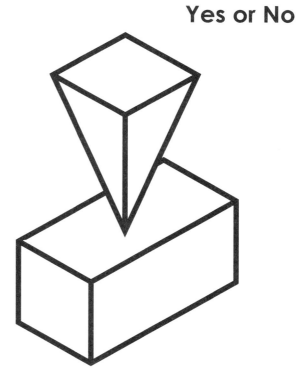

What else can you do?

Solid Combinations

What can you do with these?

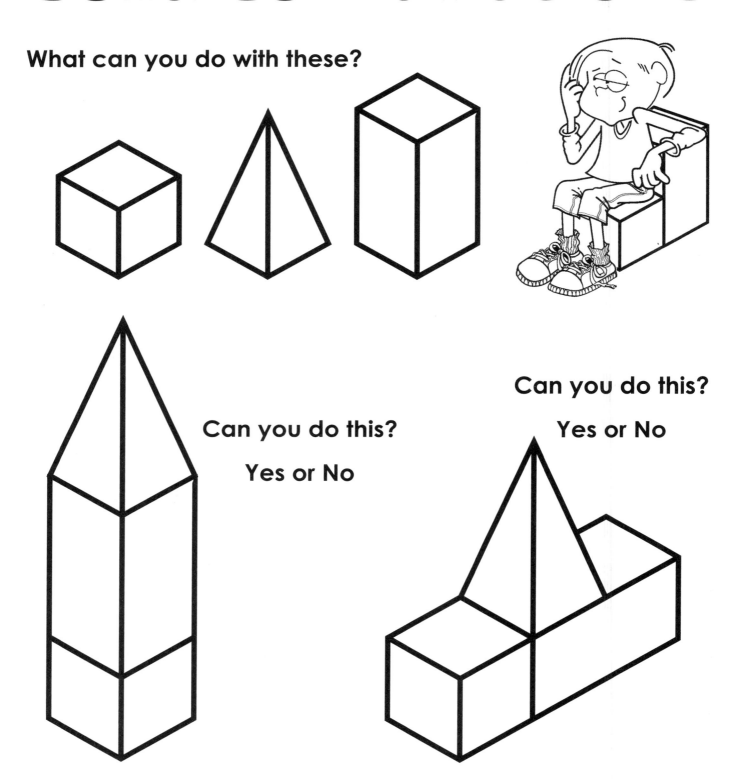

Can you do this?

Yes or No

Can you do this?

Yes or No

What else can you do?

Solid Combinations

What can you do with these?

Can you do this?

Yes or No

Can you do this?

Yes or No

What else can you do?

Characters in Position

Topic
Location/Position

Key Question
How can you model the positions of characters in a story?

Learning Goal
Students will explore location and relative position using characters from a familiar story.

Guiding Documents
NRC Standard
- *The position of an object can be described by locating it relative to another object or the background.*

*NCTM Standards 2000**
- *Describe, name, and interpret relative positions in space and apply ideas about relative position*
- *Describe, name, and interpret direction and distance in navigating space and apply ideas about direction and distance*
- *Find and name locations with simple relationships such as "near to" and in coordinate systems such as maps*

Math
Geometry
  positional words

Integrated Processes
Observing
Comparing and contrasting
Relating
Communicating

Materials
The Three Billy Goats Gruff (see *Management 1*)
Characters (see *Management 3*)
Cardboard
Scissors
File folders or card stock (see *Management 4*)

Background Information
Geometry is the study of space and shape. Spatial memory, spatial visualization and spatial reasoning allow young children to better describe, interpret, and imagine the world. Spatial sense is what helps people get around in the world and to know where various objects are in relation to themselves.

Young children need the opportunity to develop spatial sense using the relative position of objects in a variety of situations. They explore where their toys are in relation to other objects and learn to evaluate their own body movements in relation to space. The language they need to describe the relative position of themselves and the objects in their world must be practiced in a variety of contexts for effective application. These positional words include *in/out, on/off, before/after, in front/behind/between, first/last, over/under, left/right,* and others.

This activity uses a familiar story to practice the concept of relative position. While listening to a scenario, the students manipulate pictures of the characters to show the action in the story. Through multiple experiences, students become comfortable with the language needed to describe relative position.

Management
1. Select a version of *The Three Billy Goats Gruff* to read to the students. There are many versions available. See *Curriculum Correlation* for several possibilities.
2. Using two chairs and a length of cardboard, construct a bridge to use as the class acts out the story.

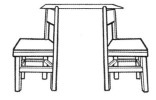

3. Duplicate one set of characters for each pair of students.
4. Each pair of students needs a file folder, a sturdy piece of card stock, or an index card to make a bridge for *Part Two*.

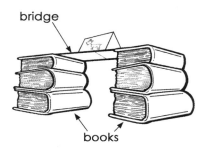

GETTING INTO GEOMETRY 171 © 2010 AIMS Education Foundation

Procedure

Part One
1. Invite two students to come to the front of the class. Introduce some of the relative position vocabulary by having the students act out your instructions. Have one student stand beside the other student, in front of the other student, to the right of and behind the other student, etc.
2. Read the story of *The Three Billy Goats Gruff* and talk about the positional words from the story.
3. Set up the bridge as described in *Management 2*.
4. Select four students to play the roles of the troll and the three billy goats. Read the story again and have them act it out as you read. Be sure students do not try to climb on top of the cardboard bridge. They can set their hands on the bridge to represent the billy goat crossing over.

Part Two
1. Tell the students that they will now all have a chance to act out parts of the story at their desks with a partner.
2. Distribute a set of characters and scissors to each pair of students and have them cut the pictures apart. Have them fold along the dashed lines and make "tents" so their characters will stand upright.

3. Give students file folders or card stock pieces and show them how to build a bridge using stacks of books (see *Management 4*).
4. Read the following instructions and tell students to move their characters to the locations described. After all the characters are in position, ask students to describe their locations relative to each other and the bridge. Repeat using various scenarios until students are comfortable with the positional words.

- Place the small billy goat under the bridge.
- Place the troll on top of the bridge.
- Place the big billy goat on the left side of the bridge.
- Place the medium size billy goat on the right side of the bridge.
- Which goat is to the right of the troll? ...the left of the troll? ...under the troll?

- Line your billy goats up from shortest to tallest on top of the bridge.
- Place the troll to the right of the bridge.
- Where is the medium size billy goat? Which billy goat is on its left? ...on its right? Where are the goats in relation to the troll?

- Place the small billy goat beside the bridge on the left.
- Place the troll under the bridge.
- Place the big billy goat beside the bridge on the right.
- Place the medium sized billy goat on top of the bridge.
- Where is the big billy goat? ...the troll? ...etc.

---

5. Use chart paper or the board to list some of the positional words used in the scenarios. Invite a student to make up a scenario that uses one or more of the words while the other students use their characters to act it out.

Connecting Learning
1. Where did the troll live? [under the bridge]
2. Who walked over the bridge? [all three billy goats]
3. How did you decide which side was to the left of the bridge? ...to the right of the bridge?
4. When the tree billy goats line up from shortest to tallest, how would you describe where the medium sized one is located? [in between the tallest and shortest]
5. When in your every day lives do you use words like above, below, beside, etc.?
6. Why are positional words important to know?

Curriculum Correlation
Applegate, Ellen. *The Three Billy Goats Gruff, A Norwegian Folktale*. Scholastic, Inc. New York. 1992.

Carpenter, Stephen. *The Three Billy Goats Gruff*. Harper Festival. New York. 1998.

Galdone, Paul. *The Three Billy Goats Gruff*. Houghton Mifflin. New York. 2001.

Kliros, Thea. *Three Billy Goats Gruff*. HarperFestival. New York. 2003.

* Reprinted with permission from *Principles and Standards for School Mathematics*, 2000 by the National Council of Teachers of Mathematics. All rights reserved.

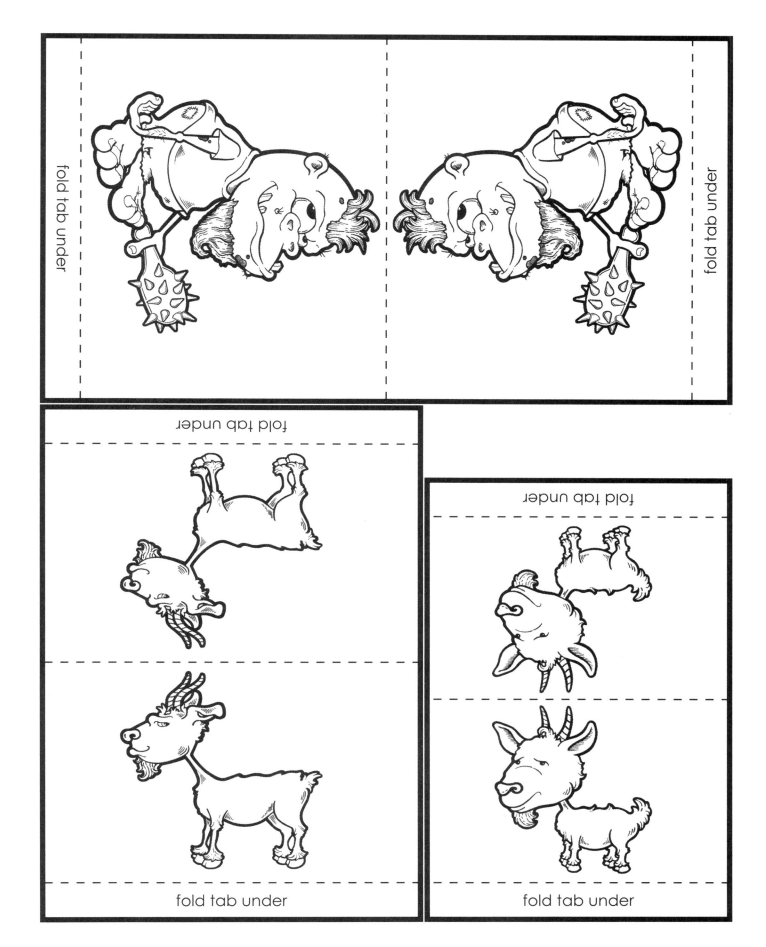

SHAPES ON LOCATION

Topic
Location/Position

Key Question
How can we make a map of shape town based on the descriptions of where each shape goes?

Learning Goals
Students will:
- place shapes on the board based on descriptions of their locations relative to other shapes, and
- describe the locations of shapes relative to other objects.

Guiding Documents
NRC Standard
- *The position of an object can be described by locating it relative to another object or the background.*

NCTM Standards 2000*
- *Describe, name, and interpret relative positions in space and apply ideas about relative position*
- *Describe, name, and interpret direction and distance in navigating space and apply ideas about direction and distance*
- *Find and name locations with simple relationships such as "near to" and in coordinate systems such as maps*

Math
Geometry
    positional words
    2-D shapes

Integrated Processes
Observing
Comparing and contrasting
Applying

Materials
Colored shapes (see *Management 1*)
Tape
Solution page (see *Management 3*)

Background Information
To describe location and position, we use positional words like *between*, *next to*, *above*, *below*, *left*, *right*, etc. These words provide different information about location, some more specific than others. To say that the trash can is next to the desk is less clear than to say that the trash can is to the right of the desk. This activity will give students the opportunity to practice applying positional words and gain confidence in using them correctly.

Management
1. Copy the page of shapes onto four colors of paper (red, yellow, blue, and green). Laminate the shapes before cutting them out.
2. Put a loop of tape or a piece of double-stick tape on the back of each shape. Arrange the shapes on a table close to the board.
3. Make a copy of the solution to display using a projection device and color the shapes as indicated.

Procedure
1. Draw a square on the board that is about two feet on each side. Explain to students that this is the boundary of shape town.
2. Tell them that they will be putting the residents in shape town according to your instructions.
3. Ask for a volunteer to come up and select the blue rectangle from the set of shapes. Instruct him or her to place the blue rectangle at the center of the town. (The orientation of the rectangle does not matter.)
4. Continue to invite students up one at a time to follow the instructions for creating shape town. (See *Instructions for Creating Shape Town*.)
5. Once all of the shapes are on the board, ask students to describe the locations of various shapes. For example, the yellow triangle is below the red square, to the left of the blue rectangle, above the yellow rectangle, and to the right of the green square.
6. Display the solution. Ask students to compare this map of shape town with the one they created on the board. Have them identify any differences and explain why they might have occurred.

GETTING INTO GEOMETRY   © 2010 AIMS Education Foundation

Connecting Learning

1. What are the words you use when you want to describe the location of an object? [next to, above, below, left, right, between, etc.]
2. Are some of these words better at describing location than others? Why? [*next to* is less specific than *to the right or left of,* etc.]
3. How did our map on the board compare to the official map? Why do you think it was not exactly the same?
4. In our map, what shapes are to the right of the yellow triangle? [blue rectangle, green circle, green triangle] Which of those shapes is closest to the yellow triangle? [blue rectangle] Which is furthest? [green triangle]
5. Describe the location of the red square in at least three different ways.
6. How would you describe the location of your desk in the classroom?
7. How would you describe the location of the office at the school?
8. What are you wondering now?

Extensions

1. Have groups of students use sets of shapes to create their own maps of shape town. Allow them to give the instructions for the other students to follow to recreate the map on the board.
2. Create other scenarios using animals, letters, magnets, etc., instead of shapes.

* Reprinted with permission from *Principles and Standards for School Mathematics*, 2000 by the National Council of Teachers of Mathematics. All rights reserved.

SHAPES ON LOCATION

Instructions for Creating Shape Town

1. Place the blue rectangle in the center of the town (the square).
2. Place the red triangle above the blue rectangle.
3. Place the green circle to the right of the blue rectangle.
4. Place the yellow square below the green circle.
5. Place the red rectangle to the right of the yellow square.
6. Place the yellow triangle to the left of the blue rectangle.
7. Place the green triangle above the red rectangle.
8. Place the red square to the left of the red triangle.
9. Place the yellow rectangle below the yellow triangle.
10. Place the red circle above the green triangle.
11. Place the green rectangle to the left of the red square.
12. Place the yellow circle below the yellow square.
13. Place the blue circle to the left of the yellow rectangle.
14. Place the blue square above the red triangle.
15. Place the green square between the green rectangle and the blue circle.
16. Place the blue triangle above the green rectangle.

Solution

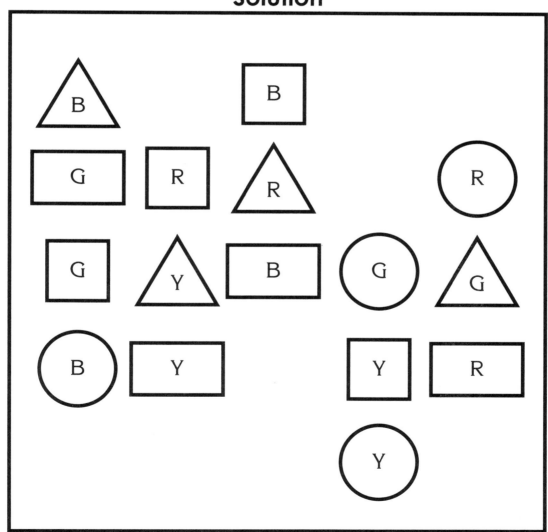

GETTING INTO GEOMETRY © 2010 AIMS Education Foundation

SHAPES ON LOCATION

Copy these shapes onto four colors of paper—red, yellow, blue, and green. Laminate the shapes and cut them out.

GETTING INTO GEOMETRY　　　178　　　© 2010 AIMS Education Foundation

Putting Shapes in Their Place

Topic
Location/Position

Key Questions
1. How can you arrange shapes on a picture according to directions given by your teacher and by a classmate?
2. How can you give directions that describe the locations of shapes on your picture?

Learning Goals
Students will:
- arrange shapes on a picture according to instructions given by the teacher,
- place shapes on their pictures in locations of their choosing,
- describe the locations of these shapes to another student using positional words, and
- see if the other student was able to position the shapes correctly based on the description given.

Guiding Documents
NRC Standard
- *The position of an object can be described by locating it relative to another object or the background.*

NCTM Standards 2000*
- *Describe, name, and interpret relative positions in space and apply ideas about relative position*
- *Describe, name, and interpret direction and distance in navigating space and apply ideas about direction and distance*
- *Find and name locations with simple relationships such as "near to" and in coordinate systems such as maps*

Math
Geometry
    positional words
    2-D shapes

Integrated Processes
Observing
Comparing and contrasting
Communicating
Identifying

Materials
Crayons
Scissors
Tape
File folders for dividers (see *Management 1*)
Playground pages (see *Management 3* and *4*)
Transparency film
Colored markers
Overhead projector (see *Management 5*)

Background Information
This activity requires students to identify shapes and colors, listen carefully to directions, and give clear instructions all while practicing the use of positional words. Not only do students follow the instructions given by the teacher, they must also take the role of the teacher themselves and give instructions to another student.

Management
1. If possible, use file folders to make dividers that can be placed between students' desks in *Part Two*. This will allow students to face each other, but not see each other's papers. Lids from paper boxes, shoeboxes and lids, poster board, or cereal boxes would also work.
2. Make enough copies of the page of shapes so that each student can have one strip of six shapes (two triangles, two rectangles, and two squares). Cut them into strips before distributing them to the students.
3. Make a large playground scene for each student. Copy the two-page version of the playground, cut along the solid line, and tape the two pages together to make one large picture.
4. Copy a set of shapes and the one-page version of the playground page to be used on a projection system. Cut apart the shapes and color them so that you have a red and an orange triangle, a yellow and a green rectangle, and a blue and a purple square.
5. If you do not have an overhead projector or document camera, you will have to modify the procedure to suit the available technology.

GETTING INTO GEOMETRY   © 2010 AIMS Education Foundation

Procedure

Part One

1. Distribute a strip of shapes, scissors, and crayons to each student. Have them color each of the six shapes a different color so that they have one red and one orange triangle, one yellow and one green rectangle, and one blue and one purple square. Instruct them to cut apart the shapes.
2. Distribute the large playground scenes you created to all students. Spend some time looking at the picture and identifying the parts and their relation to each other. For example, the bench is *to the right of* the tree. The slide is *above* the merry-go-round, and so on.
3. Tell students that they will be playing a game where they try to put their shapes on the picture exactly the same way as you do. But, they will not be able to see what you are doing. They will have to listen carefully to the directions you will give them.
4. Read the following instructions to the class, giving them time to complete the instructions after each step. You may repeat the instructions, but do not allow any questions for clarification. As you read each instruction, put the correct shape in the specified location on the overhead projector, but do not turn the projector on.
   - Put the red triangle to the left of the bench.
   - Put the green rectangle under the middle swing.
   - Put the orange triangle under the slide.
   - Put the blue square on the merry-go-round.
5. Turn on the projector so that students can see the correct arrangement of the shapes. Repeat the instructions and point out the correct location of each shape on the picture. Have students discuss whether or not they selected the correct shapes and put them in the same places as you did.
6. Have students clear their playgrounds and tell them that you have another challenge for them—this one using all six of the shapes. Read the following instructions and place your shapes on the darkened projector as you read:
   - Put the purple square on the bottom of the slide.
   - Put the green rectangle on the bench, close to the tree.
   - Put the orange triangle on top of the tree.
   - Put the red triangle to the left of the merry-go-round.
   - Put the blue square under the far left swing.
   - Put the yellow rectangle on the top of the slide.
7. Repeat the process of comparing students' pictures to your picture and discuss any problems that students had with putting the shapes in the correct locations.

Part Two

1. Explain that students are going to have the chance to play the game again, but this time, they will work in pairs, and they will get to give the directions to a partner.
2. Have students get into pairs and set up a divider between each pair. Tell them that it is important to not look at each other's papers during this activity.
3. Select one student in each pair to go first and have those students place at least four shapes on their pictures in any locations they choose.
4. Once students are ready, have them give instructions to their partners the way you gave instructions to the class. Remind them to use positional words to describe the exact locations of their shapes.
5. When students have completed their instructions, have students remove the divider and compare their papers.
6. Tell the students to reverse roles and repeat the process.
7. Close with a time of class discussion and sharing.

Connecting Learning

1. Did you use all of the correct shapes on the first try? Were they in the right locations? If not, why not?
2. Was it harder to place the shapes when you were using all six? Why or why not?
3. Was your partner able to correctly place the shapes on his/her paper based on your instructions?
4. Were you able to place your shapes on the paper based on your partner's instructions?
5. Was it easier for you to give instructions or to place your shapes? Why?
6. Why is it important to learn how to use words that describe position and location?

Extensions

1. Have students draw their own scenes on which to place the shapes.
2. Add additional shapes for students to identify and place.

* Reprinted with permission from *Principles and Standards for School Mathematics*, 2000 by the National Council of Teachers of Mathematics. All rights reserved.

GETTING INTO GEOMETRY 181 © 2010 AIMS Education Foundation

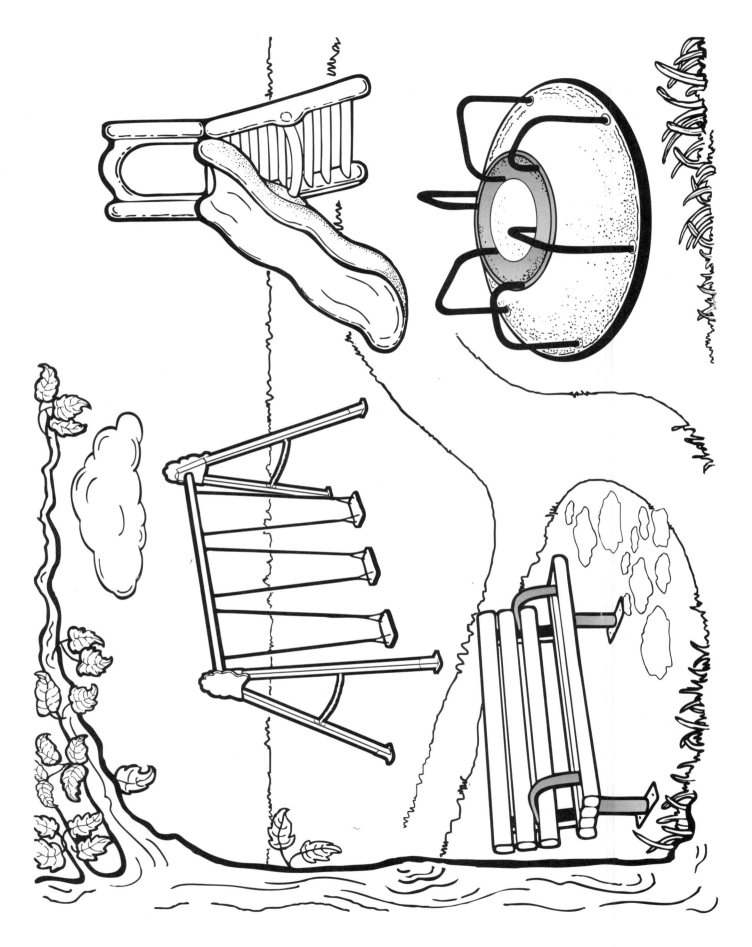

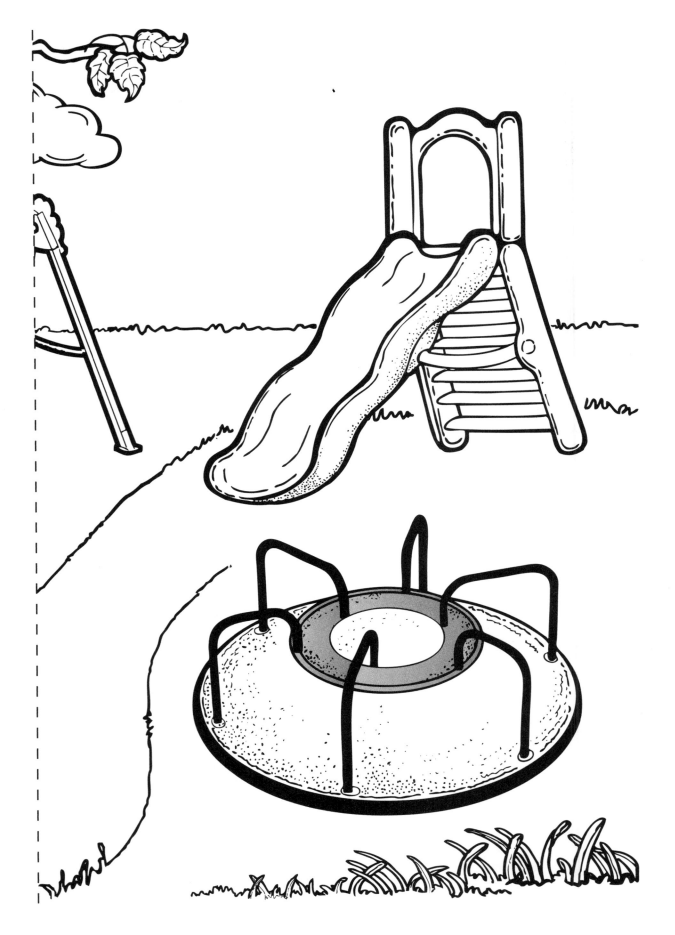

GETTING INTO GEOMETRY 184 © 2010 AIMS Education Foundation

The Tinker's Toy Store

Topic
Location/Position

Key Question
In what order are the toys arranged in Tinker's Toy Store?

Learning Goal
Students will use logical thinking skills and positional and ordinal words to describe Tinker's Toy Store.

Guiding Documents
Project 2061 Benchmarks
- *Numbers and shapes can be used to tell about things.*
- *Letters and numbers can be used to put things in a useful order.*
- *Numbers can be used to count things, place them in order, or name them.*

NRC Standard
- *The position of an object can be described by locating it relative to another object or the background.*

NCTM Standards 2000*
- *Describe, name, and interpret relative positions in space and apply ideas about relative position*
- *Describe, name, and interpret direction and distance in navigating space and apply ideas about direction and distance*
- *Find and name locations with simple relationships such as "near to" and in coordinate systems such as maps*

Math
Geometry
   positional words
Ordinal numbers
Logical thinking

Integrated Processes
Observing
Classifying
Comparing and contrasting
Communicating

Materials
For each group:
   12" x 18" construction paper
   tape
   set of 24 toy pictures
   Mr. & Mrs. Tinker figures (see *Management 2*)
   pennies
   crayons or colored pencils
   scissors
   shopping cart page

Background Information
    Positional words help us to orient ourselves and identify things in the world around us. These descriptive words include things like *over, under, to the right of, to the left of, between, behind, in front*, etc. This activity provides a playful context in which students can practice using these positional words and work on developing their spatial sense.

Management
1. This activity is divided into two parts. *Part One* involves students in organizing a toy store using positional words, directional words, and ordinal numbers. *Part Two* incorporates different positional words as students load toys into a shopping cart.
2. To construct the Tinker figures, copy them onto card stock, cut them out, and tape a penny to the tab on each figure. Fold the tab under so that the figures will stand.

Procedure

Part One

1. Distribute the Tinker figures, page of toys, scissors, and crayons or colored pencils to each group. Have the groups color the figures and toys. Instruct them to cut the page of toys into strips so that they have one strip of each kind of toy (i.e., a strip of six cars, a strip of six bears, etc.). Be sure they do NOT cut the individual toys apart.
2. Distribute the construction paper to each group. Direct students to place the construction paper on the table in front of them so that the long edge is closest to them. Tell them that this paper will represent the toy store. Explain that the front of the toy store is the side of the paper closest to them, and the back is the side farthest from them.
3. Have them place Mrs. Tinker to the right side of the toy store and Mr. Tinker behind the toy store.
4. Invite the students to name each of the toys the Tinkers have in their store. Tell them that you will read them the directions that the Tinkers will use for organizing the four aisles of their toy store. Direct them to place their toys in their toy stores according to the directions. Read the first set of directions.
5. After the students have had time to arrange their toys in the toy store, tell them that you will read the directions again so that they can check their work.
6. Ask the students where they put the aisle of teddy bears. Discuss the placement of the toys using ordinal numbers (first, second, third, and fourth).
7. Ask the students what toy is behind the balls. ...between the blocks and the balls. ...in front of the toy cars, etc.
8. Direct the students to move Mrs. Tinker to stand in front of the toy store and Mr. Tinker to stand on the left side of the toy store. Select another set of directions and continue the procedure of toy store organizing, emphasizing the use of positional words and ordinal numbers.
9. Allow time for the students to construct and describe their toy stores to each other using positional words and ordinal numbers.

Part Two

1. Tell the students that it is time to go shopping for some toys. Give each group a shopping cart page and explain that they will be using it while they shop for toys. Instruct them to cut apart the strips of toys so that they now have 24 individual pictures instead of four long strips.
2. Have them position Mr. and Mrs. Tinker next to each other, behind the toy store. Allow them time to set up the toys in their toy store in any order they choose. Ask them how many toys of each kind they have [6] and how many toys there are altogether [24]. Begin a story line about Mr. and Mrs. Tinker selecting toys to give away to the children in the hospital.
3. Inform them that they are to pick the toys and put them in the shopping cart according to the directions you will read.
4. Read the first set of directions. Ask students questions such as, What toys are in the middle of your shopping cart? [4 blocks, 3 toy cars] Where are the teddy bears in relation to the ball? [above]
5. Continue reading the directions and asking questions that deal with the placement of the toys in the shopping cart.

Connecting Learning

1. How did you know where to place the toys in the toy store?
2. Were there any directions that were confusing? Explain.
3. Was it easier for you to organize aisles in the toy store or to put toys in the shopping cart? Why?
4. Mr. and Mrs. Tinker begin to walk around their toy store. They want to stop to rest on the right side of the toy store. Place the Tinkers where they want to pause to rest.
5. Place Mr. Tinker in back of Mrs. Tinker. Describe Mrs. Tinker's position. [She's in front of Mr. Tinker.]
6. The teddy bears are between the balls and the blocks. Describe the location of the balls in relation to the teddy bears. [The balls are next to (above/below, to the right/left of) the teddy bears.]

Extensions

1. Allow students to arrange the toys in their toy stores following patterns such as AABBCCDD, ABCDABCD, etc.
2. Ask students to construct story problems about shopping for the toys. If appropriate, have them write the number sentences.

---

* Reprinted with permission from *Principles and Standards for School Mathematics*, 2000 by the National Council of Teachers of Mathematics. All rights reserved.

The Tinker's Toy Store
Mr. and Mrs. Tinker

The Tinker's Toy Store

GETTING INTO GEOMETRY 188 © 2010 AIMS Education Foundation

The Tinker's Toy Store
Clues for Part One

The blocks are in an aisle at the **front** of the toy store.

The toy cars are in an aisle at the **back** of the toy store.

The aisle of balls is right **behind** the blocks.

Put the aisle of teddy bears where you think they should go.

The teddy bears are in the **front** of the toy store.

The toy cars are **behind** the teddy bears.

The balls are in the **back** of the toy store.

The blocks are **between** the toy cars and the balls.

The teddy bears and the blocks are found **between** the balls and the cars.

The toy cars are in the **front** of the toy store.

The teddy bears are right **behind** the toy cars.

Put the balls and blocks where you think they go.

The blocks are in the **first** aisle at the **front** of the toy store.

The balls are in the **third** aisle.

The teddy bears are **between** the **first** and **third** aisles.

Put the toy cars where you think they should go.

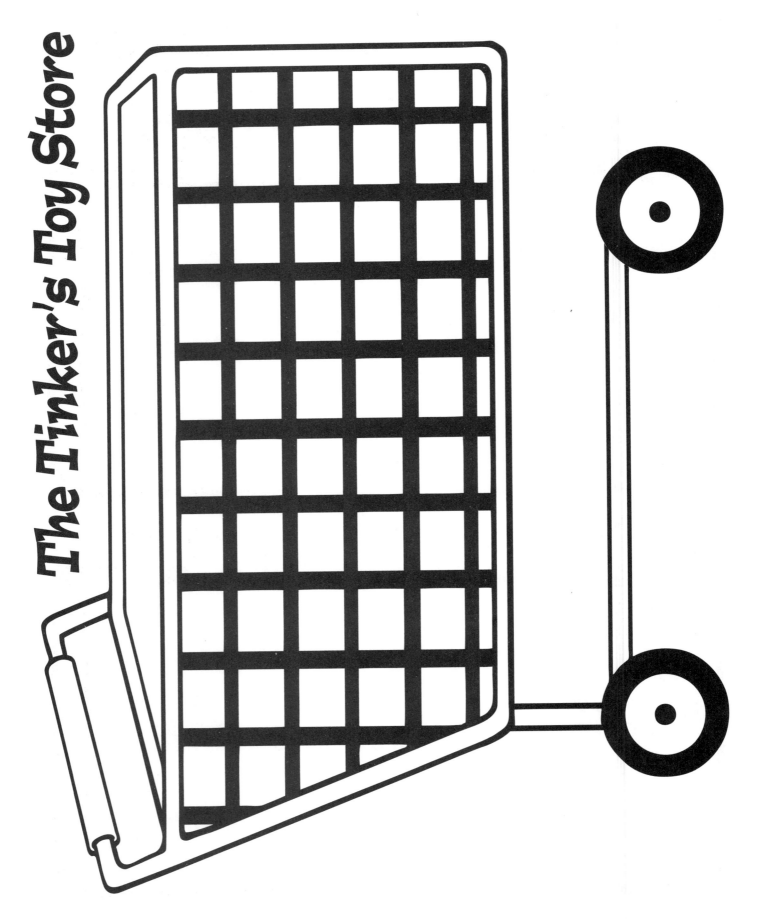

Shape Snapshots

Topic
Location/Position

Key Question
What shapes were you shown and where were they located?

Learning Goals
Students will:
- briefly look at a selection of shapes and try to recall what they saw, and
- use positional words to describe the relative positions of the shapes.

Guiding Documents
NRC Standard
- The position of an object can be described by locating it relative to another object or the background.

NCTM Standards 2000*
- Create mental images of geometric shapes using spatial memory and spatial visualization
- Describe, name, and interpret relative positions in space and apply ideas about relative position

Math
Geometry
  positional words
  2-D shapes

Integrated Processes
Observing
Comparing and contrasting
Communicating

Materials
Projection device (see *Management 1*)
Pattern blocks (see *Management 2*)
2 file folders (see *Management 3*)

Background Information
What is the configuration of three dots on a domino? Could you mentally picture the white dots diagonally positioned on the black tile? If so, you have spatial memory for that domino tile.

Spatial memory helps us draw a square or identify a triangle. It helps us to remember where the forks are in the silverware drawer without looking at the silverware tray. Visually handicapped persons use spatial memory to navigate their homes. Rapid-fire texters know the keyboards on their cell phones.

Spatial memory can be developed through purposeful experiences. In this activity, students will challenge their spatial memory of shapes as they get quick glimpses of various pattern blocks. Shape terminology and positional words will be used in context.

Management
1. Use a projection device (overhead projector, document camera, etc.) to show the pattern blocks to the students. If your students are not accountable for all of the pattern block shapes, select those that are appropriate for them to identify by name.
2. Pattern blocks are available from AIMS (item number 4250).
3. To prevent students from viewing the actual pattern blocks, use the two file folders to hide them from view.

Procedure
1. Position the file folders to form a screen so that students cannot see the pattern blocks. Place a pattern block on the projector. Turn it on for one second and then turn it off. Ask the students to identify the shape they saw. Ask them where the shape was located on the screen (top, bottom, left, right). After students have discussed the shape and its position, turn the projector on to confirm or refute their responses. Remove the pattern block and turn the projector off.
2. Position another pattern block on the projector. Follow the same procedure, allowing students to confirm or refute their responses.
3. When students are comfortable identifying one pattern block at a time, repeat the process using two and then three pattern blocks. Positional words will vary with each situation (next to, on top of, beside, etc.). Increase the exposure time as you increase the number of pattern blocks being displayed.
4. Continue the process until six pattern blocks are used. Include situations in which the same shape is used multiple times. For example: three triangles, a hexagon, and two squares.

GETTING INTO GEOMETRY   © 2010 AIMS Education Foundation

Connecting Learning
1. What shape was easiest for you to remember?
2. Was there something you did to help you remember which shapes you saw? What was it?
3. How many shapes could you easily remember?
4. What other shapes could we include?
5. What other objects could we use on the projector?

Extensions
1. Give students a set of pattern blocks so they can display the shapes and their positions shown on the projection.
2. Provide paper pattern block cutouts so students can make a record of what they saw.
3. Using multiple pattern blocks, remove a shape or change its position and ask students to identify what is different.

* Reprinted with permission from *Principles and Standards for School Mathematics*, 2000 by the National Council of Teachers of Mathematics. All rights reserved.

Getting Around Geoville

Topic
Location/Position

Key Question
How can you find your way on the map by listening to directions?

Learning Goals
Students will:
- identify two-dimensional and three-dimensional shapes on a map, and
- navigate specific routes on the map based on the directions given.

Guiding Documents
NRC Standards
- *The position of an object can be described by locating it relative to another object or the background.*
- *An object's motion can be described by tracing and measuring its position over time.*

NCTM Standards 2000*
- *Recognize, name, build, draw, compare, and sort two- and three-dimensional shapes*
- *Describe, name, and interpret relative positions in space and apply ideas about relative position*
- *Describe, name, and interpret direction and distance in navigating space and apply ideas about direction and distance*
- *Find and name locations with simple relationships such as "near to" and in coordinate systems such as maps*

Math
Geometry
- positional words
- 2-D shapes
- 3-D solids

Integrated Processes
Observing
Comparing and contrasting
Identifying
Applying

Materials
Map page
Teddy Bear Counters
Sheet protectors
Dry erase markers
Paper towels

Background Information
Location, relative position, direction, and distance are key components of spatial awareness that children begin to develop from an early age. This activity provides students the opportunity to practice applying some of the terms that accompany these concepts such as *right, left, between, in front of, behind,* and so on, in the context of a map. The added challenge is for students to maintain the perspective of the bear that they are moving on the map rather than their own perspective. This can be facilitated by turning the map so that the bear's perspective and the student's perspective are the same. The landmarks on the map are all two- and three-dimensional shapes, so shape identification is also a component of the experience.

Management
1. This activity is divided into two parts. These parts can be done on the same day or on different days.
2. For *Part One*, each student needs a Teddy Bear Counter or similar three-dimensional piece that stands upright and has an identifiable front and back. Teddy Bear Counters are available from AIMS (item number 1924).
3. For *Part Two*, place the maps in sheet protectors and have students use dry erase markers to trace the routes. Each route can be erased before the next one is traced.
4. Students must be able to identify right and left in order to do this activity.
5. No directions should cause students (or the bears) to go off the map.
6. Optional instructions are included for both parts of this activity that use the cardinal directions (north, south, east, west).

GETTING INTO GEOMETRY © 2010 AIMS Education Foundation

Procedure

Part One

1. Ask students if they have ever looked at a map. Have them identify the kinds of things you find on maps. [place names, roads, boundaries, etc.] Ask them to describe some of the ways we use maps. [to find where a country (state, city, etc.) is located; to find our way from one location to another; to find our way around a place (such as an amusement park); etc.]
2. Tell students that you have a street map of a place called Geoville. In Geoville all of the buildings are either geometric shapes or solids. The map shows all of the streets in town and all of the buildings.
3. Distribute the map to students. Have them identify each of the shapes and solids. Direct them to pay careful attention to the shapes and solids that might be easily confused, such as the rectangle and rectangular solid or the cube and the square or the triangle and the cone.
4. Once students are comfortable with the names of all the shapes and solids, distribute a Teddy Bear Counter to each student. Explain that they will be guiding their bears through the town based on directions that you will read them. They must do everything from the bear's perspective. If you say to turn right, it must be the bear's right. If you say to go left, it must be the bear's left.
5. To gain some practice, have students place their bears on the maps in the intersection between the square, rectangular solid, cone, and rectangle. Instruct them to turn their bears to face the square and the rectangular solid. Show them how to turn their papers so that the bear's back is toward them and it is facing the same direction they are. Explain that it may be helpful to turn the map so that the bear's back is always to them. That way the bear's right is the same as their right.
6. Ask students to identify the shape that is ahead and to the right. [rectangular solid] Have them identify the shape that is ahead and to the left. [square] Instruct them to turn their bears to the left, and ask them what shapes the bear now sees in front of it. [rectangle and square] Repeat this practice as needed until you are confident that students are able to navigate from the bear's perspective.
7. Tell students that they are now ready to follow some directions to move their bears around Geoville. Read the first set of instructions, one line at a time, and circulate among the students to check that they are able to correctly follow the directions.
8. Repeat with the remaining instructions and/or develop instructions of your own.

Part Two

1. Distribute sheet protectors, paper towels, and dry erase markers. Have students place their maps inside the sheet protectors. Show them how to make a mark on the map with the dry erase marker and then erase it using the paper towel.
2. Explain that they are going to have a different challenge using the map this time. Instead of moving a Teddy Bear Counter on the map, they are going to use the dry erase marker to trace a path based on your instructions.
3. Have students practice following your instructions by telling them to start between the cube and the hexagon and trace a path that goes straight up to the rectangle. Have them erase this line and practice with a few other simple instructions until you are confident they are ready for the more complex instructions.
4. Read the first set of instructions, one line at a time, and circulate among the students to see how they are doing. When you get to the end of a set of instructions, have students compare their maps with their neighbors. Did everyone trace the same path?
5. Invite several students to describe the path by identifying the shapes that were seen along the way, etc.
6. Repeat with the remaining instructions and/or develop instructions of your own.

Connecting Learning

1. Why do we use maps? Have you ever used a map? How did you use it?
2. What things did you need to know to follow the instructions on the map of Geoville? [left and right, directional and positional words, etc.]
3. Were there any instructions that were hard for you to follow? Explain.
4. Did you think it was harder to move the bear around town or to trace the path that someone walked? Explain.
5. If your teddy bear is standing with its back to the triangle, what shapes could be across the street? [rectangle, cone, cylinder]
6. What is ahead when the cone is to your bear's left and the circle is to your bear's right? [rectangular solid]
7. If Tommy is walking with the cone on his right and the rectangular solid on his left, what shape is ahead on the right? [circle]

Extensions

1. Modify the maps to include different shapes and solids.
2. Have students trace a path of their own choosing on the map and describe that path to a classmate in such a way that the other student can correctly replicate the path.

---

* Reprinted with permission from *Principles and Standards for School Mathematics*, 2000 by the National Council of Teachers of Mathematics. All rights reserved.

Getting Around Geoville

Part One Instructions

Easier
Start with the oval on your left and the cube on your right.
Turn left at the first corner.
What is on your left? [oval]
What is on your right? [cylinder]

Start with the triangle on your left and the cylinder on your right.
Walk to the next block.
What is on your left? [cone]
What is on your right? [circle]
What is in front of you? [rectangular solid]

Start with the square on your right and the rectangular solid on your left.
Turn left at the first corner.
Walk along that street to the last block.
Describe what you see on your left and right. [hexagon and cube]

More Difficult
Start between the cylinder and the circle.
Walk toward the rectangle.
Turn the corner so the cone is on your right.
Walk to the next corner.
What is ahead and to your left? [square]

Start between the hexagon and the cube.
Walk toward the corner with the circle.
Turn left at the corner and pass between the cube and the circle.
Turn right at the next corner.
As you walk to the rectangle, what shapes do you pass on the left? [cylinder and triangle]

Start between the triangle and the rectangle.
Walk toward the cone.
Turn right and pass between the cone and the triangle.
At the corner, turn to face the rectangular solid.
Which way did you turn? [left]

Part Two Instructions

Teddy took a walk around town.
He started with the square on his left and the rectangle on his right.
Make an X where Teddy started.
He passed by the cone and the circle.
He turned and walked to the cylinder.
Where did Teddy walk?

Tammy took a walk around town.
She started with the oval on her right and the cylinder on her left.
Make an X where Tammy started.
She walked past the circle and turned a corner.
She saw the circle on her left.
She turned another corner and saw the cone on her right.
Where did Tammy walk?

Theodore took a walk around town.
He started with his back to the rectangular solid.
He walked toward the cylinder and passed the circle on his left.
Make an X where Theodore started.
He turned a corner and walked past the cone on his right.
He turned another corner and the cone was still on his right.
He walked toward the square, then turned a corner and was back where he started.
Where did Theodore walk?

Part One Instructions, Cardinal Directions

Draw a compass rose in one corner of the map page.

Start with the oval to the west of you and the cube to the east of you.
Go north.
Turn east at the second intersection.
What is to the north? [cone]
What is to the south? [circle]

Start with the cube to the west of you and the hexagon to the east of you.
Go north.
When you pass the cone to the west, turn east.
What is to the north? [square]
What is to the south? [rectangular solid]

Start with your back to the rectangle.
The triangle should be ahead to the west and the cone ahead to the east.
Go south two blocks.
Turn to the west.
What is to the north? [cylinder]
What is to the south? [oval]

Part Two Instructions, Cardinal Directions

Terry took a walk around town.
She started with the square on her left and the rectangle to her right.
Make an X where Terry started.
She headed south and turned west at the third intersection.
She traveled one block west then turned north.
She traveled one block north and then turned west.
Where did Terry walk?

Tony took a walk around town.
He started with the triangle to the west and the cone to the east.
Make an X where Tony started.
He headed north and turned east at the first intersection.
He walked one block east and turned south at the next intersection.
He walked south until he ran out of road.
Where did Tony walk?

 # Getting Around Geoville

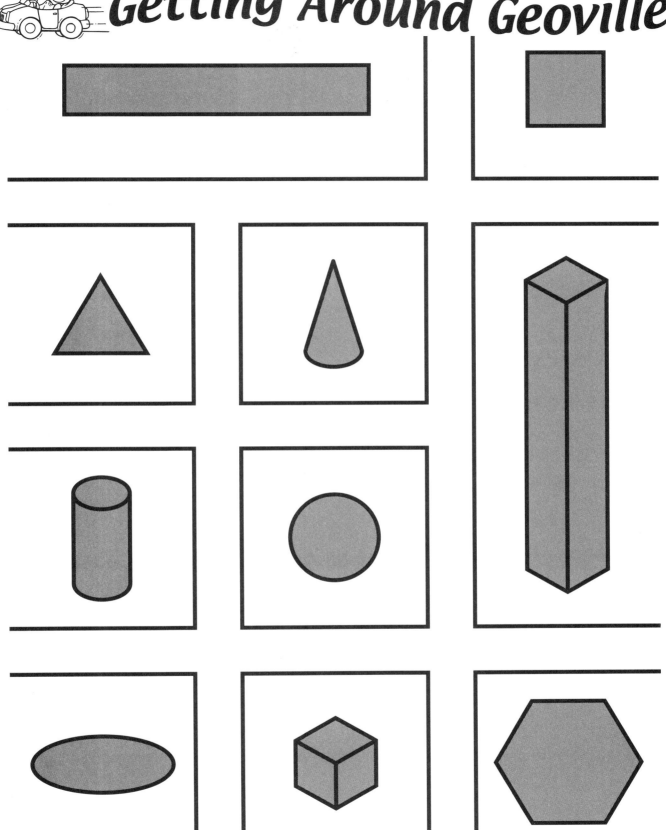

Spin and Win

Topic
2-D shapes

Key Question
How can playing a game help us learn about shapes?

Learning Goals
Students will:
- play a game to explore characteristics of shapes; and
- identify circles, triangles, and rectangles, including squares.

Guiding Documents
Project 2061 Benchmarks
- *Numbers and shapes can be used to tell about things.*
- *Shapes such as circles, squares, and triangles can be used to describe many things that can be seen.*

*NCTM Standards 2000**
- *Recognize, name, build, draw, compare, and sort two- and three-dimensional shapes*
- *Describe attributes and parts of two- and three-dimensional shapes*
- *Recognize and represent shapes from different perspectives*

Math
Geometry
    2-D shapes
        characteristics
        identification

Integrated Processes
Observing
Comparing and contrasting
Communicating

Materials
For each group of students:
    game spinner
    game card for each student in the group
    game markers (see *Management 3*)

For the class:
    paper fasteners
    small paper clips
    eyelets (see *Management 2*)
    projection device (see *Management 4*)

Background Information
Young children need multiple opportunities to use the vocabulary of geometry in order for them to internalize the terms. This playful practice encourages the proper use of vocabulary and reinforces the basic attributes of circles, squares, triangles, and rectangles. It presents the shapes in a variety of orientations.

Management
1. Copy the game cards and spinners onto card stock and laminate for extended use. You will need one game card per student and one spinner for each group of students.
2. To make spinners, place an eyelet with the wide side down in the center of the spinner page. Place the large end of a small paper clip over the eyelet, push the paper fastener through the center of the eyelet and paper, and spread the ends to secure all the pieces in place. Eyelets are available in fabric stores or craft departments.

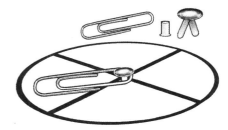

3. Any small objects that will fit in the spaces on the game cards can be used as markers. Beans, buttons, pennies, or plastic tiles will all work. Each student needs four or five markers.
4. If you do not have document camera or similar device that you can use to demonstrate the spinner, gather the students around a table to show them how to flick the paper clip.

Procedure
1. Draw a triangle on the board. Ask the class how they would describe the shape you just drew. Prompt them with questions about the number of edges it has, the number of corners, etc.
2. Draw a circle, square, rectangle, and triangle on the board. Ask the class to point to and name the shape with three edges, four equal edges, etc.
3. Explain that the students will be learning about some geometric shapes by playing a game.
4. Display the game board for the class to see. Ask students to identify each shape.
5. Demonstrate the proper use of the spinner for the class (see *Management 4*). Spin it and discuss where the spinner landed. For example, if it lands on the 3, question the students about what shape on the game board has three edges. Explain that they would then use a game marker to cover a triangle on their card. Tell the class that the object is to be the first to cover three in a row diagonally, vertically, or horizontally.
6. Allow time for students to play in small groups.
7. Introduce the second spinner. Compare the two spinners.
8. Allow time for students to play using the second spinner.
9. End with a review of the characteristics of the circle, triangle, rectangle, and square. Encourage students to share what they learned from playing the game.

Connecting Learning
1. What is a three-sided shape called? [triangle]
2. How many edges does a rectangle have? [four]
3. How can you tell if a four-sided shape is a square? [All of its edges are the same.]
4. How would you describe a circle?
5. How did playing a game help you learn about shapes?

* Reprinted with permission from *Principles and Standards for School Mathematics,* 2000 by the National Council of Teachers of Mathematics. All rights reserved.

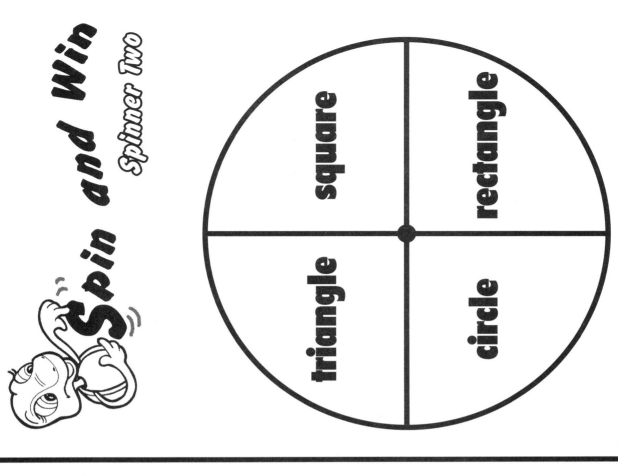

Spin and Win
Spinner Two

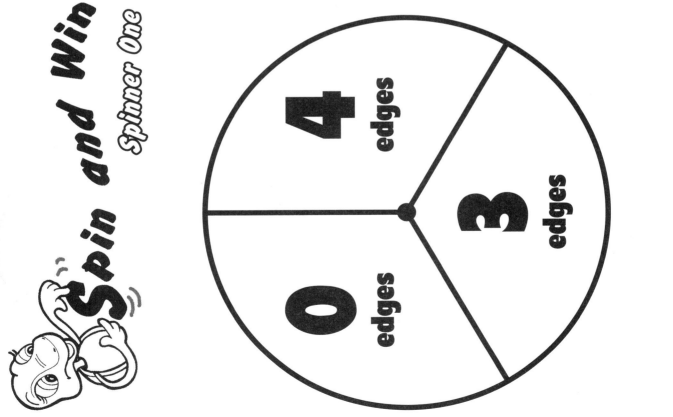

Spin and Win
Spinner One

GETTING INTO GEOMETRY © 2010 AIMS Education Foundation

What's in My Pocket?

Topic
2-D shapes/3-D solids

Key Question
What is the shape in my pocket?

Learning Goal
Students will listen to clues about geometric shapes and use them to identify mystery shapes.

Guiding Document
NCTM Standards 2000*
- *Describe attributes and parts of two- and three-dimensional shapes*
- *Recognize, name, build, draw, compare, and sort two- and three-dimensional shapes*

Math
Geometry
    2-D shapes
    3-D solids

Integrated Processes
Observing
Communicating
Applying

Materials
Clue cards
Large sticky notes
Shapes and solids cards, one set per student
Craft sticks, 10 per student
Scissors
Glue or tape

Background Information
    The riddles in this activity were designed to help students pay close attention to the characteristics of shapes and solids. It will also reinforce vocabulary and communication skills.

Management
1. Copy clue cards (the shirts with pockets) onto card stock and laminate for extended use.
2. Place a large sticky note to cover what is in the pictured pocket.
3. Copy a set of shapes and solids cards on card stock for each student.
4. The clue cards can later be placed in a center for additional practice.
5. Blank clue cards have been included so that students can create their own sets of clues for the shapes they know.

Procedure
1. Ask students to name as many geometric shapes and solids as possible.
2. Tell the students that they are going to play a riddle game called *What's in My Pocket?* Explain that you will read a clue and they will try to guess what mystery shape or solid is hidden in the pocket.
3. Give each student scissors, a set of cards, and 10 craft sticks. Assist students in cutting out the shape cards and attaching the craft sticks.

4. Read a clue and have students raise the appropriate card when they know what shape or solid is hidden in the pocket. Repeat the process until all shapes and solids have been identified.
5. If desired, provide additional time for students to create their own clue cards for additional shapes and solids they know.

Connecting Learning
1. Were some clues better than others? Explain.
2. Which was easier, creating your own or solving others' riddles? Why?
3. Did you find a riddle that had more than one solution? Explain.
4. Can the same shape have a different set of clues? Explain.

* Reprinted with permission from *Principles and Standards for School Mathematics*, 2000 by the National Council of Teachers of Mathematics. All rights reserved.

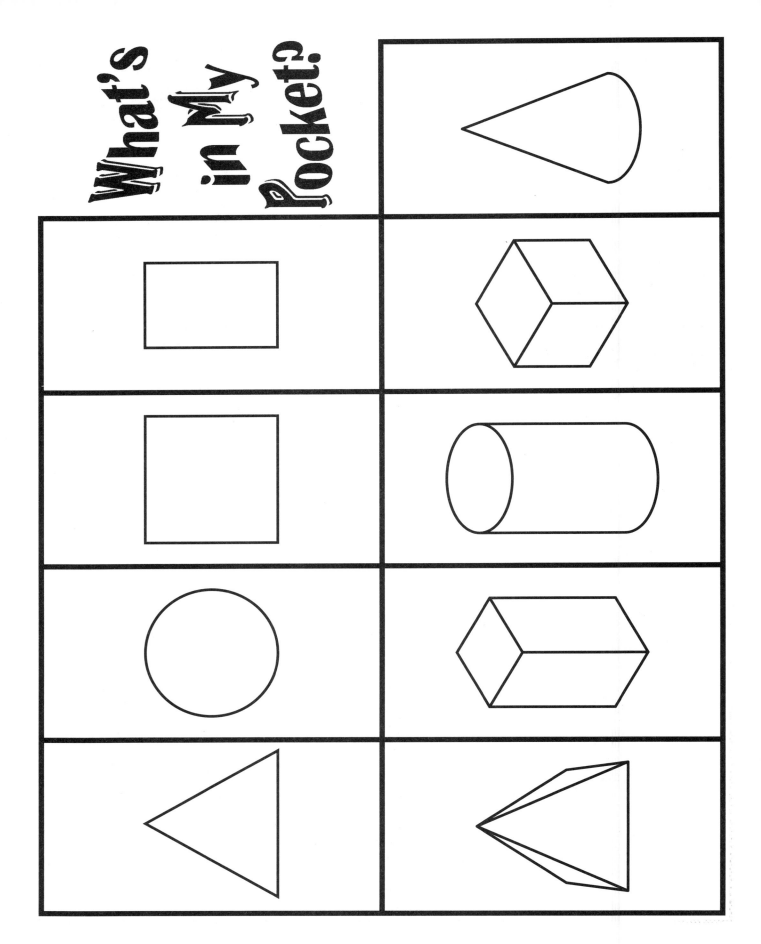

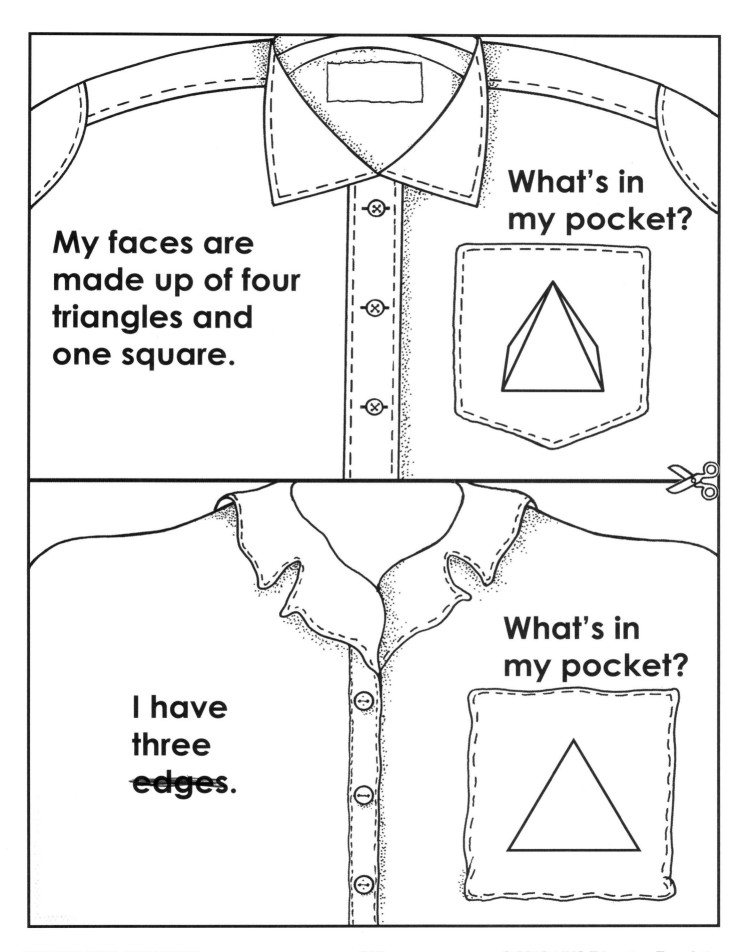

Boat Builders

Topic
Geometry

Key Question
What do the different shapes represent on our glyphs?

Learning Goal
Students will use a variety of 2-D shapes to represent various preferences on a glyph.

Guiding Documents
Project 2061 Benchmarks
- *Numbers and shapes can be used to tell about things.*
- *Shapes such as circles, squares, and triangles can be used to describe many things that can be seen.*

NCTM Standards 2000*
- *Pose questions and gather data about themselves and their surroundings*
- *Represent data using concrete objects, pictures, and graphs*
- *Recognize, name, build, draw, compare, and sort two- and three-dimensional shapes*

Math
Geometry
  2-D shapes
Data analysis

Integrated Processes
Observing
Comparing and contrasting
Classifying
Collecting and recording data
Interpreting data
Drawing conclusions
Communicating

Materials
Glue sticks
Crayons
Scissors
Student pages

Background Information
Glyphs are a pictorial form of data collection. Hieroglyphics were a method of writing used by ancient Egyptians and others. The word *glyph* is short for the word *hieroglyphics*. In this activity, geometric shapes have been used so that students can gather data and review shape identification. Each glyph represents a unique characteristic of information. Students construct a display of specific glyphs based on their responses to a set of questions. The questions are the key or legend to the glyphs. Once the displays are created, students are asked to analyze and interpret the data. For example, this illustration of a completed glyph represents a boy who has been to the beach, would rather build sand castles, prefers to swim a pool, and whose favorite vacation spot is the mountains.

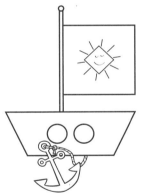

Management
1. Have several examples of picture writing available for your students to examine. Include such things as Egyptian hieroglyphics, Native American pictographs, and present-day symbols such as the handicap accessible symbol. (See *Internet Connections* for websites with examples.)
2. Display the legend where all students can easily see it. This can be done by enlarging it onto chart paper or by using a projection device.
3. This activity should be spread out over several days.
4. Make one copy of the student pages for each student. If possible, copy the page with the mast of the boat on card stock. Prior to teaching this lesson, complete a glyph that represents your responses to the legend questions.

GETTING INTO GEOMETRY © 2010 AIMS Education Foundation

Procedure

1. Lead your students in a discussion about picture writing and hieroglyphics. Show the students several examples and have them infer what each picture represents. Tell your students that they will be doing some picture writing to communicate information about themselves.
2. Show the students your completed glyph. Ask them what they notice about the boat. [The boat is shaped like a trapezoid; it has a triangular sail, two circles for portholes, etc.] Tell the students that the pictures represent things about you.
3. Draw the students' attention to the legend. Show them how the shape of the boat means that you have been to the beach, and that the shape of the sail means that you would rather swim in the ocean. Continue using the legend to interpret your sample boat. To reinforce writing, you may ask the students to help you write sentences that describe you, based on your picture. For example, I am a woman who has been to the beach. I like to build sandcastles. Etc.
4. Distribute a set of student pages, crayons, scissors, and a glue stick to each student.
5. Ask, "Who has been to the beach?" Direct students' attention to the boat type section of the first student page. Instruct them to circle the shape that represents their answer. Have them put an X through the shape that does not represent their answer.
6. Draw students' attention to the sail shape section of the survey. Have students circle the picture that represents where they would prefer to swim—the ocean, a lake, or a pool. Ask students to put an X on the sails that they will not be using.
7. Continue to read each question on the survey aloud. Have the students circle their responses and put Xs on the pictures that do not apply to them.
8. Allow time for the students to color the parts of their boat glyphs on the subsequent pages that correspond to the responses that were circled.
9. Distribute the page with the boat's mast and a glue stick to each student.
10. Direct the students to cut out their colored responses and to glue the pictures onto their mast page. Collect the boat glyphs.
11. Display a glyph and ask the students to use the legend to help you describe the owner of the boat. Repeat the process several times until the students are comfortable with using a legend to "read" the glyphs.

Connecting Learning

1. How many students in our class have been to the beach? How do you know that?
2. What shape sail would you have if you would rather swim in a pool? [square]
3. Do more people like to swim in pools or the ocean? How do you know?
4. What shapes are there in our glyphs?
5. How many circles are on your glyph?
6. How are our glyphs like hieroglyphics? How are they different?
7. Choose a glyph. Based on the glyph, what do you know about the student who created it?

Extension

Have students sort the glyphs and use them to create Venn diagrams and picture graphs.

Internet Connections

Hieroglyphs
http://www.greatscott.com/hiero/

Wikipedia: Petroglyphs
http://en.wikipedia.org/wiki/Petroglyph

Southwestern United States Rock Art Gallery
http://net.indra.com/~dheyser/rockart.html

* Reprinted with permission from *Principles and Standards for School Mathematics*, 2000 by the National Council of Teachers of Mathematics. All rights reserved.

Boat Builders
Legend

Have you ever gone to the beach?

Yes

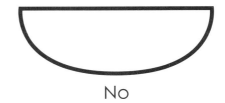
No

I'd rather...

swim in the ocean

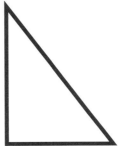
swim in a lake

swim in a pool

I am a...

girl

boy

I'd rather...

O
read a book

OO
build a sand castle

OOO
play a game

My favorite vacation spot is...

the beach

anchor on the left

the mountains

anchor in the middle

neither

anchor on the right

Boat Builders

Boat Type
Have you ever gone to the beach?

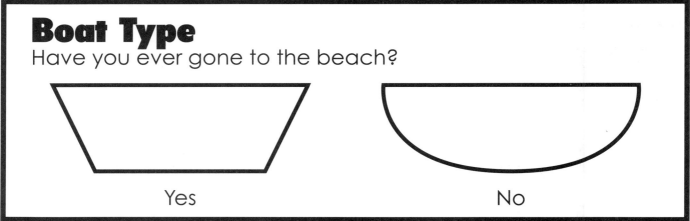

Yes   No

Sail Shape
I'd rather…

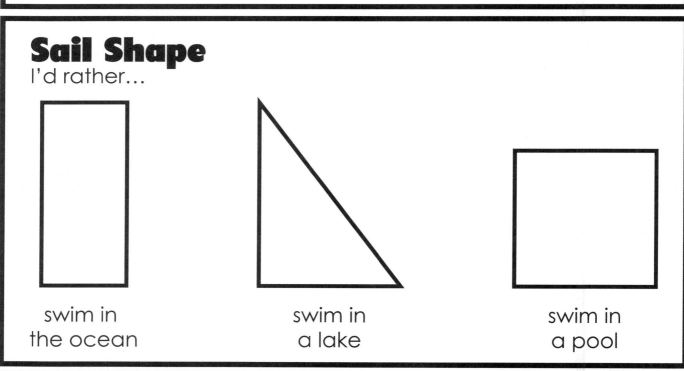

swim in the ocean   swim in a lake   swim in a pool

Portholes
I'd rather…

read a book   build a sand castle   play a game

Boat Builders

Sail Decoration

I am a...

girl

boy

Anchor Location

My favorite vacation spot is...

| the beach | the mountains | neither |

anchor on the left / anchor in the middle / anchor on the right

Boat Builders

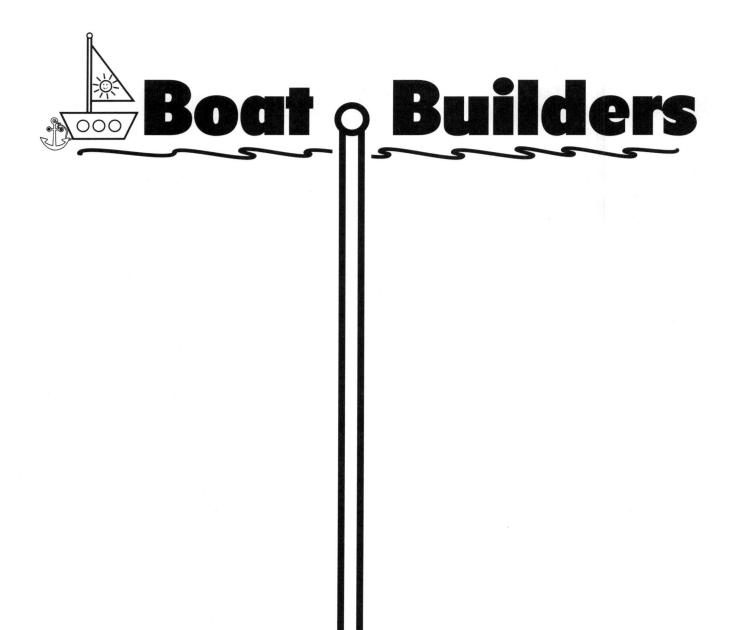

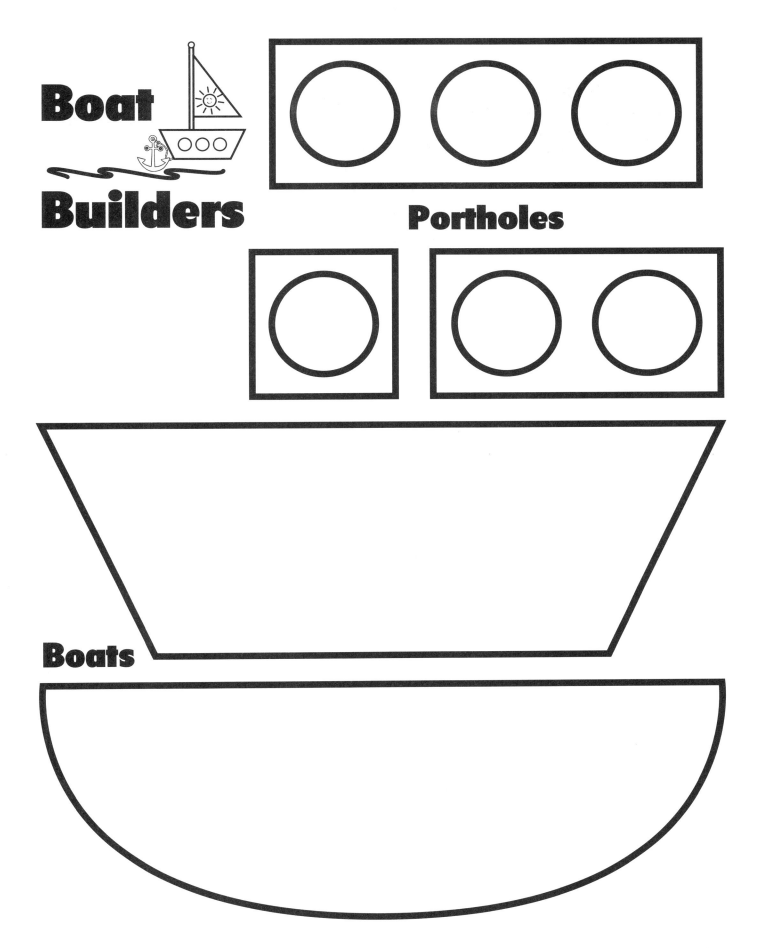

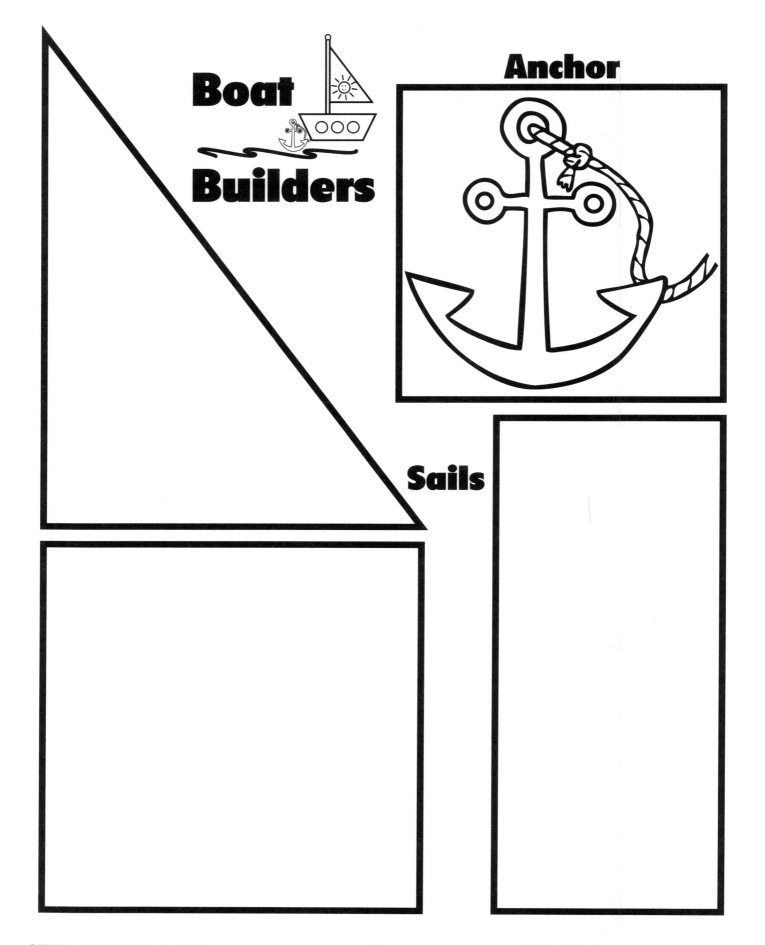

Twisting to Shapes

Purpose of the Game
Students will play a game similar to the commercial Twister® game where they will be given verbal directions to place either a foot or hand onto a specific shape.

Materials
Tape
Spinner
Eyelet
Paper clip
Paper fastener
Shape cards
Markers

Management
1. Copy the spinner onto card stock and laminate for extended use. Place the eyelet in the center of the paper. Put the small end of the paper clip over the eyelet, then place the paper fastener through the eyelet and the paper. Spread the ends of the paper fastener to secure the spinner.

2. Make six copies of each shape on card stock for each group. Copy the squares on green paper, the circles on red paper, the rectangles on blue paper, and triangles on yellow paper (or color the shapes after copying). Create a mat by securely taping the shapes to the floor so that there is a column of squares on the far left, a column of triangles next to the squares, a column of circles next the triangle, then a column of rectangles on the far right.

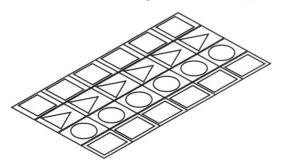

3. Divide the class into groups of three or five.
4. The game requires either two or four players. Designate the extra person in each group as the referee. The referee is not considered a player; during the game, the referee will spin the spinner, call out the moves, and monitor the game play.

Rules
1. Players take off their shoes, and position themselves on the mat according to the number of players, as explained below.
   For 2-players:
   Players face each other from opposite ends of the mat.
   For 4-players:
   Players position themselves on all four side of the mat.
2. The referee in each group spins the spinner, then calls out the body part and the shape that the spinner points to. For example, the referee may call out: "Right hand, triangle" All players, at the same time, must try to place the called-out body part on a vacant triangle. If the called-out hand or foot is already on the shaped called, they must try to move it to another of the same shape.
3. There can never be more than one hand or foot on any one shape. If two or more players reach for the same shape, the referee must decide which player got there first. The other player(s) must find another vacant shape.
4. Never remove your hand or foot from a shape unless the referee directs you to after a spin. Exception: You may lift a hand or foot to allow another hand or foot to pass by, as long as you announce it to the referee beforehand and replace it on its shape immediately afterward. If all six shapes are already covered, the referee must spin again until a different shape can be called out.
5. As soon as a player falls or touches the mat with an elbow or knee, the player is eliminated. The last player left in the game is the winner!

Twisting to Shapes

Twisting to Shapes

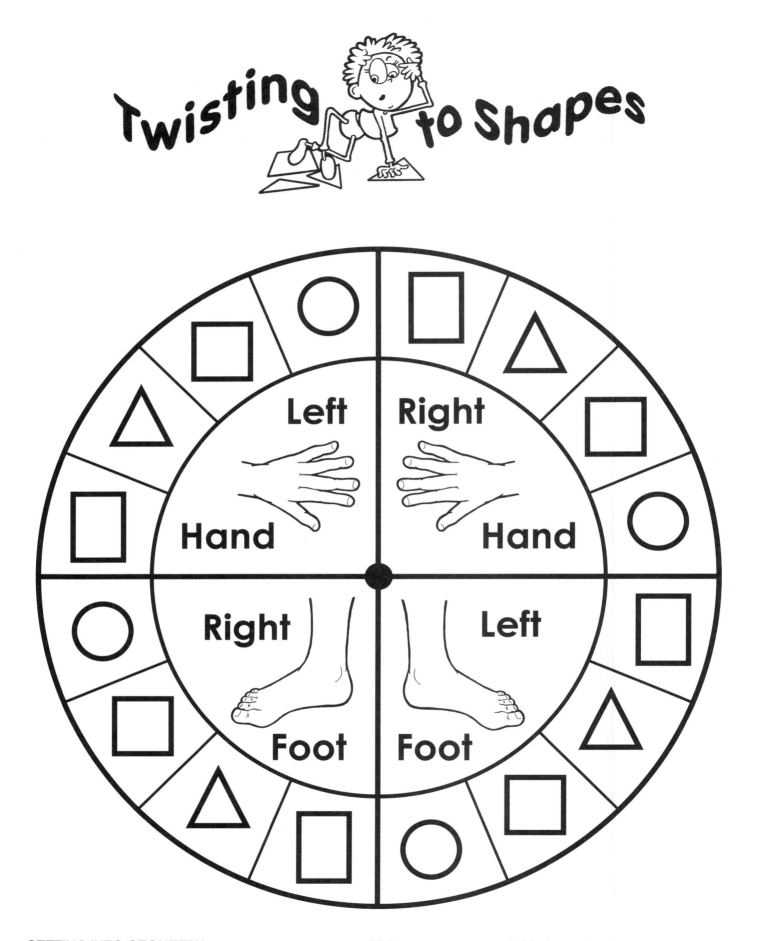

Eyes-on Geometry

Purpose of the Game
Students will determine what object has been selected by a classmate based on color, shape, and location clues.

Management
1. Play this game with at least two players. It is suggested that the entire class play.

Rules
1. One player looks around the room and mentally selects an object that can be seen by all the players.
2. The player says, "I spy with my little eye," and then gives a shape and color description of the object, such as, "a red square."
3. The other players take turns asking yes or no questions that will assist them in guessing what the object is.
4. The player who correctly guesses the selected item picks the next object. (Or, all players take turns in a set order.)

Symmetry Pair-O

Purpose of the Game
Students will match cards to their symmetric halves on a game board in an attempt to be the first to cover all the shapes in a row, column, or diagonal.

Materials
For each student:
  game boards
  set of cards, cut apart
  small cup or container for cards

Management
1. This game can be played in groups of two to four players.
2. Copy a game board on card stock for each student.
3. Copy one set of cards on transparency film for each student. (When the cards are copied on transparency film, players can align the dashed lines on the card to the lines on the game board to see the entire shape.) Cut apart the cards and place them in a small cup or container from which the student can draw.

Rules
1. A player begins by drawing a card from his/her set of cards and matching it with its symmetric half on the game board.
2. Play continues with players taking turns drawing cards and matching them to their symmetric halves on the game board.
3. The first player to get four in a row vertically, horizontally, or diagonally is the winner. Variation: The first player to cover all four corners is the winner.

SYMMETRY PAIR-O
Game Board

SYMMETRY PAIR-O
Cards

GETTING INTO GEOMETRY © 2010 AIMS Education Foundation

Matchmakers

Purpose of the Game
Students will form symmetric matches from their cards and try to avoid being the player to end with the unmatched card.

Materials
Matchmakers cards

Management
1. This game is for groups of three or four.
2. Copy the *Matchmakers* cards on card stock and cut them apart. Make one set for each group of students. Copying the card sets on different colors of card stock will help keep them separate. Laminate for extended use.

Rules
1. Deal the cards evenly among all players. (Some players may have more cards than others, depending on the number of players.)
2. Any cards in players' hands that form symmetric matches are placed on the table in front of the players.
3. One player begins by drawing another card from any other player. If it is a match to one of the cards in his/her hand, he/she places the symmetric pair on the table. If not, the card remains in his/her hand, and play moves to the next person.
4. Players take turns drawing cards from each others' hands until all of the cards are matched but one. The player with this card loses one point.
5. Each pair of cards is worth one point, and the player with the most points is the winner.

Matchmakers

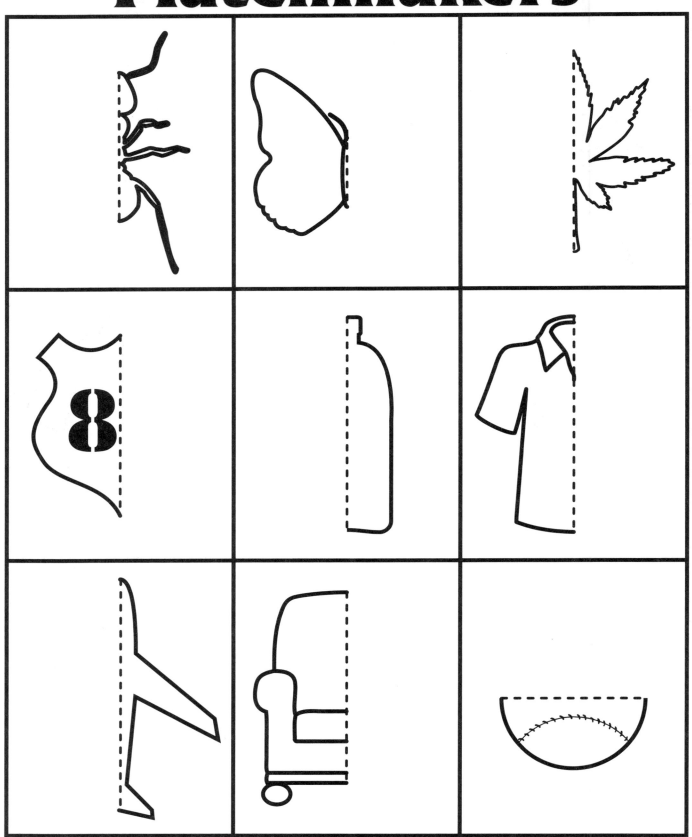

Matchmakers

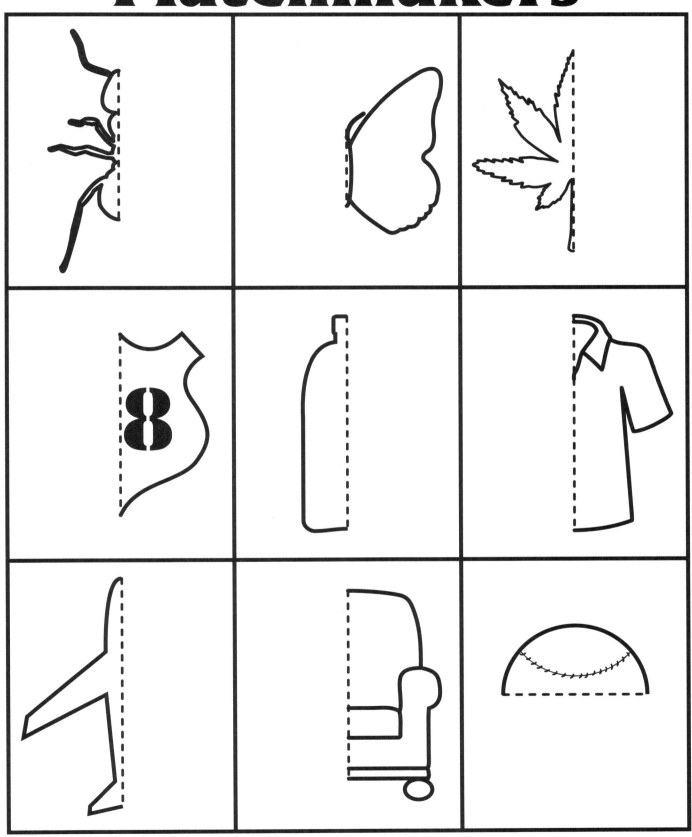

Matchmakers

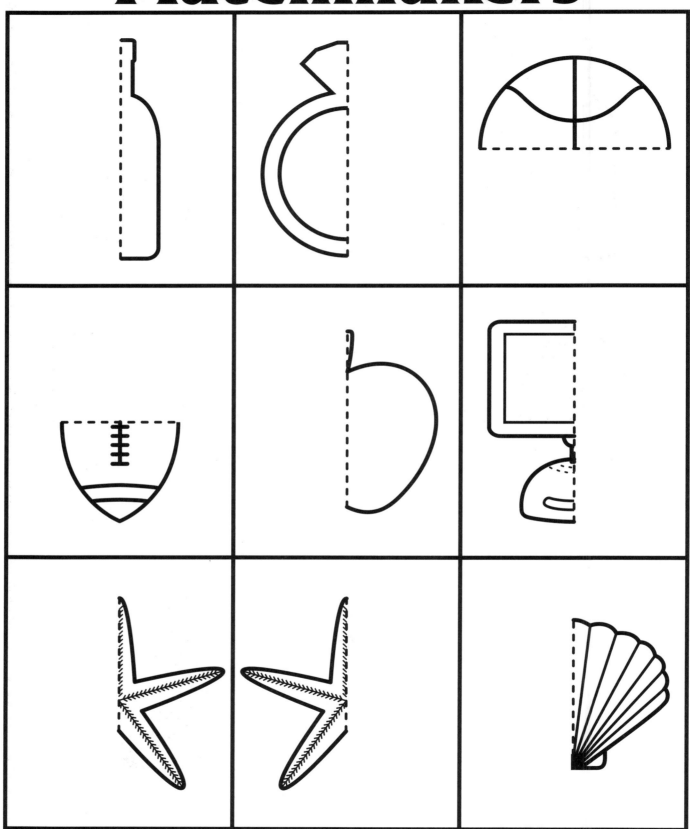

Matchmakers

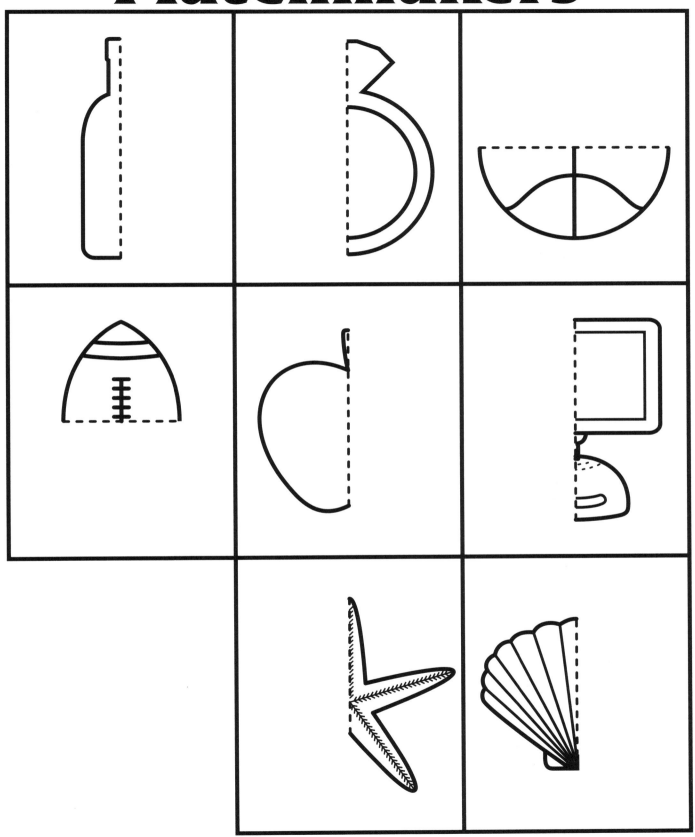

Concentrating on Shapes and Solids

Purpose of the Game
Students will play a concentration game to practice identifying and using the names of common 2-D shapes and 3-D solids.

Materials
Game cards

Management
1. Copy the game cards onto card stock and laminate for extended use. You will need one set of cards for each game you make. It is suggested that you first model the game, and then allow students to play in small groups. Mark the cards in each set with a different number, letter, or sticker, or copy them on different colors of paper. This makes it easy to sort the cards or find in which deck a stray card belongs.
2. There are 24 cards that can be used in a variety of combinations:
   - 2-D shapes to 2-D shapes—Remove the cards with words and the cards with pictures of 3-D solids.
   - 2-D shapes (and/or 3-D solids) to words—Remove one of the two squares, triangles, rectangles, and circles.
   - 2-D shapes, three-way match—Remove all of the 3-D solids and corresponding words; instead of matching only two cards, students must match three (both pictures and the word).
3. This game is for two to four players.

Rules
1. Mix up the cards. Lay them out face down in rows.
2. The first player turns over two (or three) cards. If the shape(s) and/or the word match, the player takes the cards and gets another turn. If they don't match, the cards are returned to the face-down position and it is the next player's turn.
3. Continue taking turns until all the cards are matched up.
4. The winner is the one with the most matches.

GETTING INTO GEOMETRY © 2010 AIMS Education Foundation

Circle	**Cone**
Square	**Cylinder**
Triangle	**Cube**
Rectangle	**Rectangular Solid**

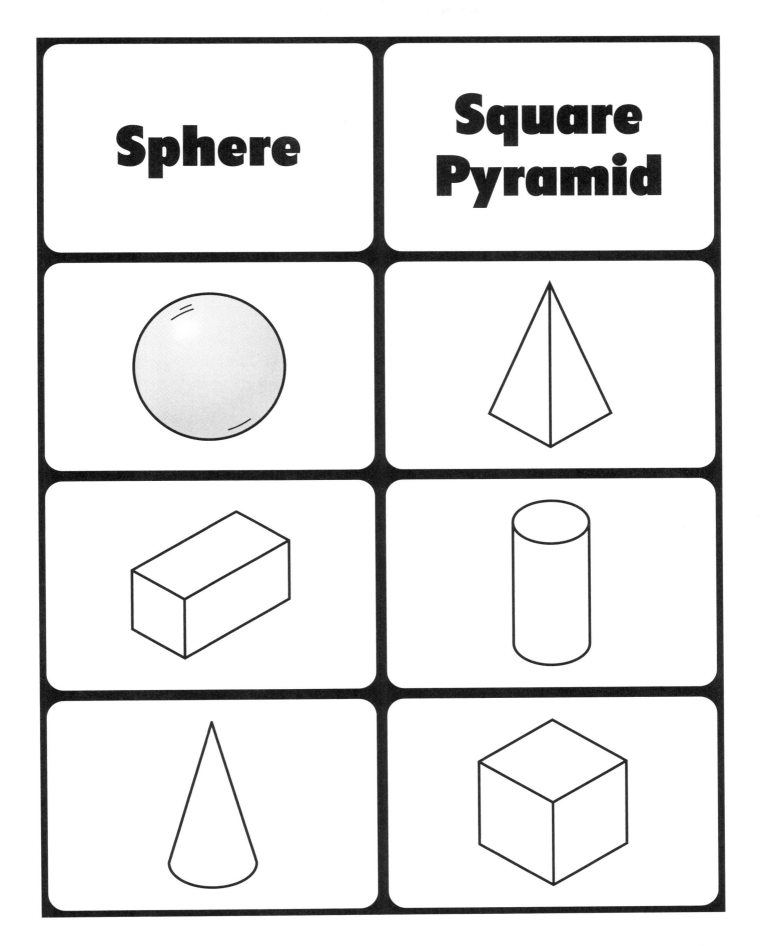

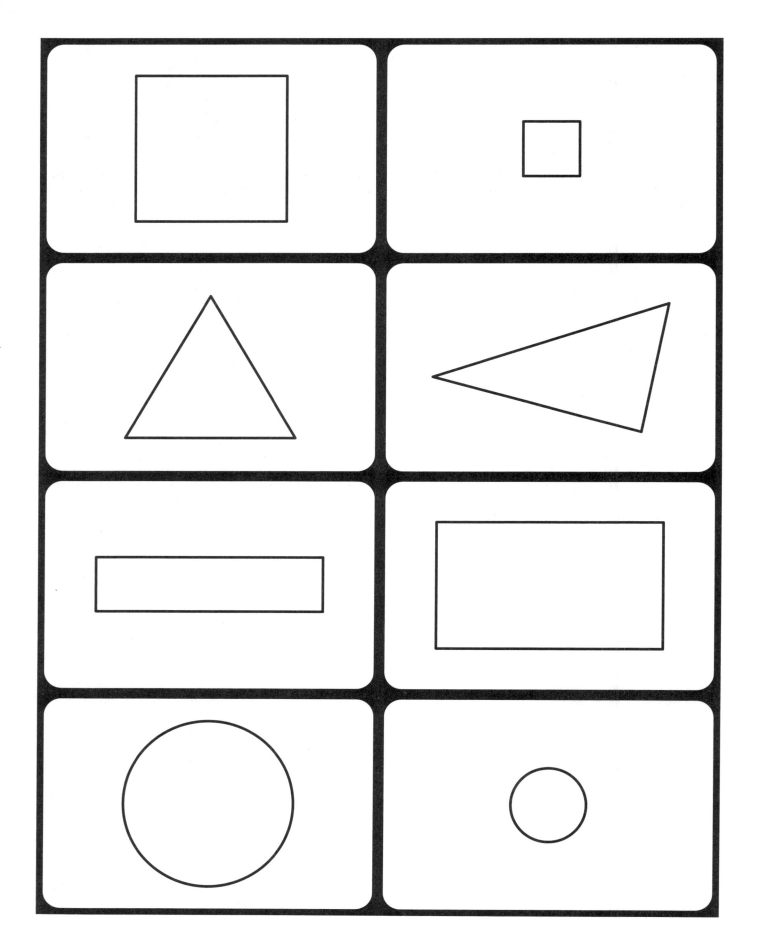

WHO HAS? Shapes and Solids

Purpose of the Game
Students will participate in a game where they will identify 2-D shapes and/or 3-D solids.

Materials
Who Has? cards (1-20 or 21-40)

Management
1. There are two sets of cards included. The first set of 20 cards uses 2-D shapes and the other uses 3-D solids. If you have more than 20 students, some will need to share cards.
2. Copy the cards onto card stock. You may wish to copy each set on a different color to help keep them separate. Color the 2-D shapes on the first set as indicated.
3. Laminate the cards for extended use and cut them apart.

Rules
1. Each student (or pair of students) gets one card.
2. Any student begins by reading his or her card aloud. (For example, "I have a red square. Who has a green rectangle?")
3. The student with the answer to the card responds by reading the card aloud. (For example, "I have a green rectangle. Who has a blue triangle?")
4. The game continues until the cycle returns to the beginning card. (In this example, "I have a red square.")

Who Has? 2-D Shapes Key
1. I have a red square. Who has a green rectangle?
2. I have a green rectangle. Who has a blue triangle?
3. I have a blue triangle. Who has a yellow circle?
4. I have a yellow circle. Who has a red rectangle?
5. I have a red rectangle. Who has two green triangles?
6. I have two green triangles. Who has a blue circle?
7. I have a blue circle. Who has a purple triangle?
8. I have a purple triangle. Who has an orange square and a green square?
9. I have an orange square and a green square. Who has two yellow circles?
10. I have two yellow circles. Who has a brown triangle?
11. I have a brown triangle. Who has an orange rectangle?
12. I have an orange rectangle. Who has a purple circle?
13. I have a purple circle. Who has a yellow square on a table?
14. I have a yellow square on a table. Who has a red square under a table?
15. I have a red square under a table. Who has a blue triangle to the right of a table?
16. I have a blue triangle to the right of a table. Who has a green circle to the left of a table?
17. I have a green circle to the left of a table. Who has an orange rectangle above a table?
18. I have an orange rectangle above a table. Who has a purple triangle below a table?
19. I have a purple triangle below a table. Who has a yellow square to the left of a table?
20. I have a yellow square to the left of a table. Who has a red square?

Who Has? 3-D Solids Key
21. I have a sphere. Who has a rectangular solid?
22. I have a rectangular solid. Who has two cubes?
23. I have two cubes. Who has a cone and a cylinder?
24. I have a cone and a cylinder. Who has a sphere and a cube?
25. I have a sphere and a cube. Who has two rectangular solids?
26. I have two rectangular solids. Who has a cylinder?
27. I have a cylinder. Who has a sphere and a cone?
28. I have a sphere and a cone. Who has two spheres?
29. I have two spheres. Who has three cones?
30. I have three cones. Who has a cube and a rectangular solid?

GETTING INTO GEOMETRY © 2010 AIMS Education Foundation

31. I have a cube and a rectangular solid. Who has a cone?
32. I have a cone. Who has a cube?
33. I have a cube. Who has two cylinders?
34. I have two cylinders. Who has a cylinder and a rectangular solid?
35. I have a cylinder and a rectangular solid. Who has three spheres?
36. I have three spheres. Who has two cones?
37. I have two cones. Who has a sphere and a rectangular solid?
38. I have a sphere and a rectangular solid. Who has three cubes?
39. I have three cubes. Who has two cubes and a cone?
40. I have two cubes and a cone. Who has a sphere?

I have a red square.

Who has a green rectangle?

I have a green rectangle.

Who has a blue triangle?

I have a blue triangle.

Who has a yellow circle?

I have a yellow circle.

Who has a red rectangle?

I have a red rectangle.

Who has two green triangles?

I have two green triangles.

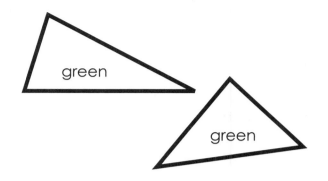

Who has a blue circle?

I have a blue circle.

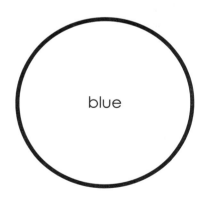

Who has a purple triangle?

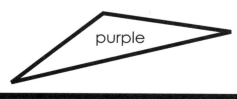

I have a purple triangle.

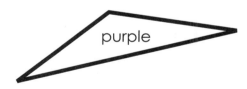

Who has an orange square and a green square?

I have an orange square and a green square.

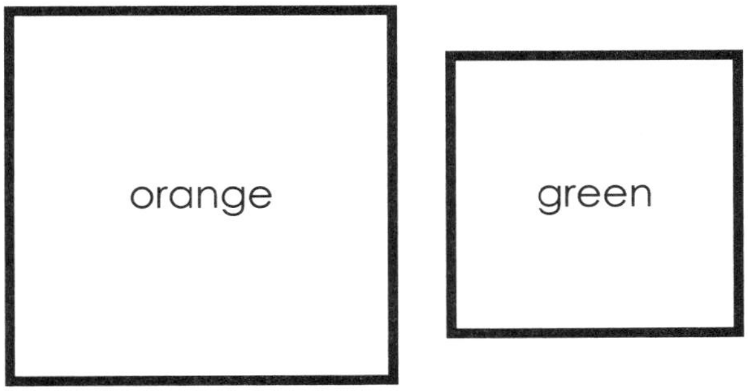

orange   green

Who has two yellow circles?

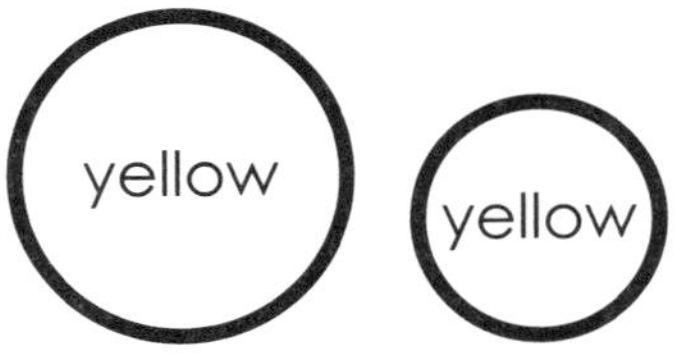

yellow yellow

I have two yellow circles.

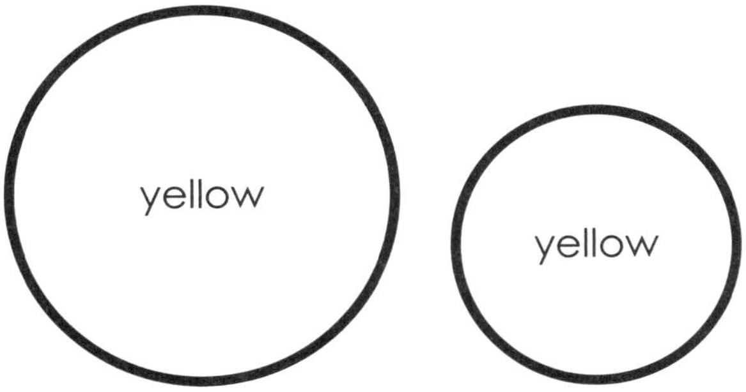

yellow yellow

Who has a brown triangle?

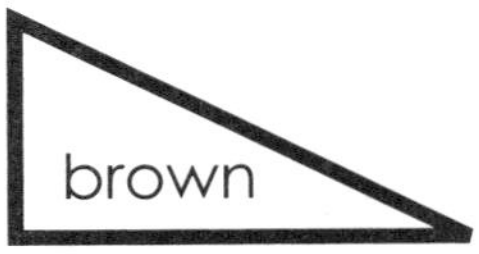

brown

I have a brown triangle.

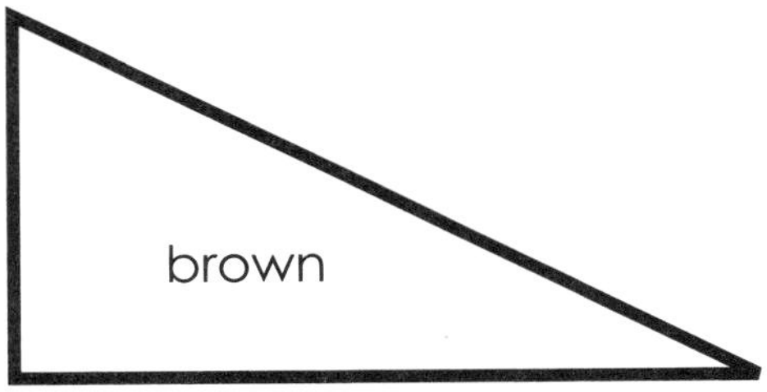

brown

Who has a orange rectangle?

orange

I have an orange rectangle.

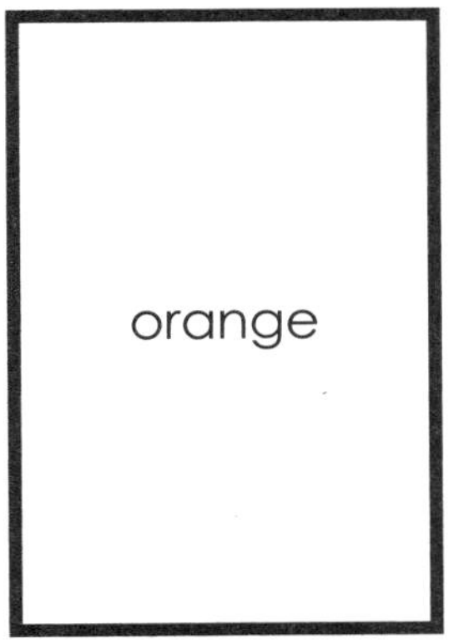

orange

Who has a purple circle?

purple

GETTING INTO GEOMETRY

I have a purple circle. 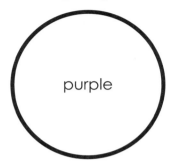 Who has a yellow square on a table? 	I have a yellow square on a table.  Who has a red square under a table? 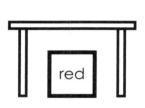
I have a red square under a table. 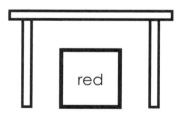 Who has a blue triangle to the right of a table? 	I have a blue triangle to the right of a table. 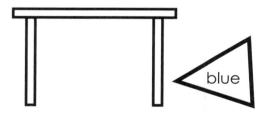 Who has a green circle to the left of a table? 

I have a green circle to the left of a table.

Who has an orange rectangle above a table?

I have an orange rectangle above a table.

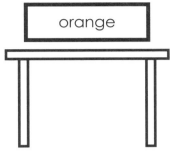

Who has a purple triangle below a table?

I have a purple triangle below a table.

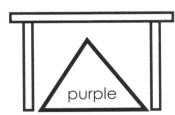

Who has a yellow square to the left of a table?

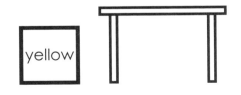

I have a yellow square to the left of a table.

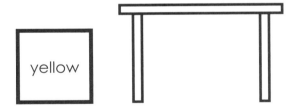

Who has a red square?

GETTING INTO GEOMETRY © 2010 AIMS Education Foundation

I have a sphere.  Who has a rectangular solid? 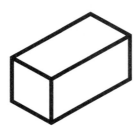	I have a rectangular solid. 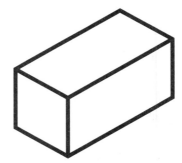 Who has two cubes?  
I have two cubes. 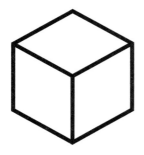  Who has a cone and a cylinder?  	I have a cone and a cylinder.  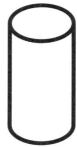 Who has a sphere and a cube?  

GETTING INTO GEOMETRY © 2010 AIMS Education Foundation

I have a sphere and a cube.

 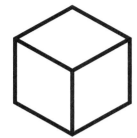

Who has two rectangular solids?

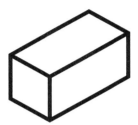 

I have two rectangular solids.

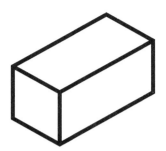 

Who has a cylinder?

I have a cylinder.

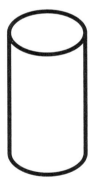

Who has a sphere and a cone?

I have a sphere and a cone.

Who has two spheres?

GETTING INTO GEOMETRY

I have two spheres.

Who has three cones?

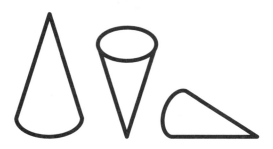

I have three cones.

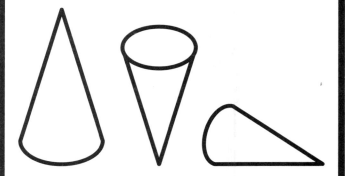

Who has a cube and a rectangular solid?

I have a cube and a rectangular solid.

 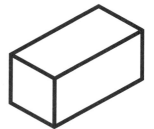

Who has a cone?

I have cone.

Who has a cube?

GETTING INTO GEOMETRY © 2010 AIMS Education Foundation

I have a cube.

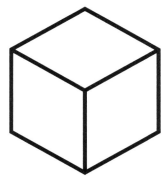

Who has two cylinders?

 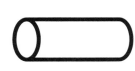

I have two cylinders.

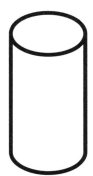

Who has a cylinder and a rectangular solid?

I have a cylinder and a rectangular solid.

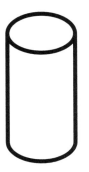 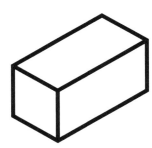

Who has three spheres?

I have three spheres.

Who has two cones?

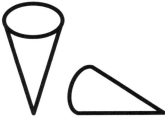

I have two cones.

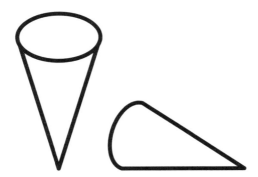

Who has a sphere and a rectangular solid?

 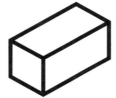

I have a sphere and a rectangular solid.

 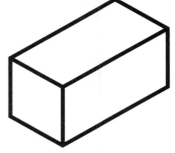

Who has three cubes?

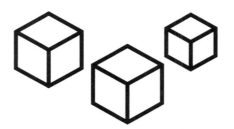

I have three cubes.

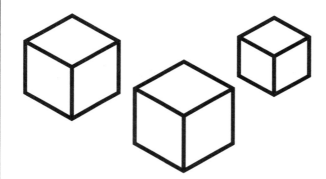

Who has two cubes and a cone?

I have two cubes and a cone.

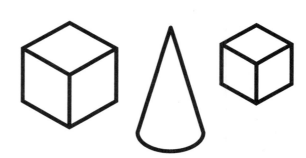

Who has a sphere?

Literature for Geometry
Kindergarten and Grade One

Applegate, Ellen. *The Three Billy Goats Gruff, a Norwegian Folktale.* Scholastic, Inc. New York. 1992.

Campbell, Kathy Kuhtz. *Let's Draw a Bear With Squares.* Powerkids Press. New York. 2004.

Carpenter, Stephen. *The Three Billy Goats Gruff.* HarperFestival. New York. 1998.

Crews, Donald. *School Bus.* HarperTrophy. New York. 1993.

Galdone, Paul. *The Three Billy Goats Gruff.* Houghton Mifflin. New York. 2001.

Jonas, Ann. *Round Trip.* Greenwillow Books. New York. 1990.

Maccarone, Grace. *Three Pigs, One Wolf, and Seven Magic Shapes.* Scholastic, Inc. New York. 1997.

Kliros, Thea. *Three Billy Goats Gruff.* HarperFestival. New York. 2003.

Murphy, Stuart J. *Let's Fly a Kite.* HarperCollins. New York. 2000.

Nelson, Robin. *Circle.* Learner Publishing Group, Inc. Minneapolis, MN. 2004.

Nelson, Robin. *Rectangle.* Learner Publishing Group, Inc. Minneapolis, MN. 2004.

Nelson, Robin. *Square.* Learner Publishing Group, Inc. Minneapolis, MN. 2004.

Nelson, Robin. *Triangle.* Learner Publishing Group, Inc. Minneapolis, MN. 2004.

Parker, Marjorie Blain. *Hello, School Bus!* Scholastic, Inc. New York. 2004.

Roth, Carol. *The Little School Bus.* North-South Books. New York. 2002.

Singer, Marilyn. *I'm Your Bus.* Scholastic Press. New York. 2009.

Thong, Roseanne. *Round is a Mooncake.* Chronicle Books. San Francisco. 2000.

Van Voorst, Jennifer. *Making Shapes.* Capstone Press. Bloomington, MN. 2003.

Whitford, Ann Paul. *Eight Hands Round: A Patchwork Alphabet.* HarperCollins. New York. 1996.

Music
The Wheels on the Bus by Raffi or Maryann Kovalski.

The AIMS Program

AIMS is the acronym for "**A**ctivities **I**ntegrating **M**athematics and **S**cience." Such integration enriches learning and makes it meaningful and holistic. AIMS began as a project of Fresno Pacific University to integrate the study of mathematics and science in grades K-9, but has since expanded to include language arts, social studies, and other disciplines.

AIMS is a continuing program of the non-profit AIMS Education Foundation. It had its inception in a National Science Foundation funded program whose purpose was to explore the effectiveness of integrating mathematics and science. The project directors, in cooperation with 80 elementary classroom teachers, devoted two years to a thorough field-testing of the results and implications of integration.

The approach met with such positive results that the decision was made to launch a program to create instructional materials incorporating this concept. Despite the fact that thoughtful educators have long recommended an integrative approach, very little appropriate material was available in 1981 when the project began. A series of writing projects ensued, and today the AIMS Education Foundation is committed to continuing the creation of new integrated activities on a permanent basis.

The AIMS program is funded through the sale of books, products, and professional-development workshops, and through proceeds from the Foundation's endowment. All net income from programs and products flows into a trust fund administered by the AIMS Education Foundation. Use of these funds is restricted to support of research, development, and publication of new materials. Writers donate all their rights to the Foundation to support its ongoing program. No royalties are paid to the writers.

The rationale for integration lies in the fact that science, mathematics, language arts, social studies, etc., are integrally interwoven in the real world, from which it follows that they should be similarly treated in the classroom where students are being prepared to live in that world. Teachers who use the AIMS program give enthusiastic endorsement to the effectiveness of this approach.

Science encompasses the art of questioning, investigating, hypothesizing, discovering, and communicating. Mathematics is a language that provides clarity, objectivity, and understanding. The language arts provide us with powerful tools of communication. Many of the major contemporary societal issues stem from advancements in science and must be studied in the context of the social sciences. Therefore, it is timely that all of us take seriously a more holistic method of educating our students. This goal motivates all who are associated with the AIMS Program. We invite you to join us in this effort.

Meaningful integration of knowledge is a major recommendation coming from the nation's professional science and mathematics associations. The American Association for the Advancement of Science in *Science for All Americans* strongly recommends the integration of mathematics, science, and technology. The National Council of Teachers of Mathematics places strong emphasis on applications of mathematics found in science investigations. AIMS is fully aligned with these recommendations.

Extensive field testing of AIMS investigations confirms these beneficial results:

1. Mathematics becomes more meaningful, hence more useful, when it is applied to situations that interest students.
2. The extent to which science is studied and understood is increased when mathematics and science are integrated.
3. There is improved quality of learning and retention, supporting the thesis that learning which is meaningful and relevant is more effective.
4. Motivation and involvement are increased dramatically as students investigate real-world situations and participate actively in the process.

We invite you to become part of this classroom teacher movement by using an integrated approach to learning and sharing any suggestions you may have. The AIMS Program welcomes you!

AIMS Education Foundation Programs

When you host an AIMS workshop for elementary and middle school educators, you will know your teachers are receiving effective, usable training they can apply in their classrooms immediately.

AIMS Workshops are Designed for Teachers
- Correlated to your state standards;
- Address key topic areas, including math content, science content, and process skills;
- Provide practice of activity-based teaching;
- Address classroom management issues and higher-order thinking skills;
- Give you AIMS resources; and
- Offer optional college (graduate-level) credits for many courses.

AIMS Workshops Fit District/Administrative Needs
- Flexible scheduling and grade-span options;
- Customized (one-, two-, or three-day) workshops meet specific schedule, topic, state standards, and grade-span needs;
- Prepackaged four-day workshops for in-depth math and science training available (includes all materials and expenses);
- Sustained staff development is available for which workshops can be scheduled throughout the school year;
- Eligible for funding under the Title I and Title II sections of No Child Left Behind; and
- Affordable professional development—consecutive-day workshops offer considerable savings.

University Credit—Correspondence Courses
AIMS offers correspondence courses through a partnership with Fresno Pacific University.
- Convenient distance-learning courses—you study at your own pace and schedule. No computer or Internet access required!

Introducing AIMS State-Specific Science Curriculum
Developed to meet 100% of your state's standards, AIMS' State-Specific Science Curriculum gives students the opportunity to build content knowledge, thinking skills, and fundamental science processes.
- Each grade-specific module has been developed to extend the AIMS approach to full-year science programs. Modules can be used as a complete curriculum or as a supplement to existing materials.
- Each standards-based module includes math, reading, hands-on investigations, and assessments.

Like all AIMS resources, these modules are able to serve students at all stages of readiness, making these a great value across the grades served in your school.

For current information regarding the programs described above, please complete the following form and mail it to: P.O. Box 8120, Fresno, CA 93747.

Information Request

Please send current information on the items checked:

___ *Basic Information Packet* on AIMS materials
___ Hosting information for AIMS workshops
___ AIMS State-Specific Science Curriculum

Name: _____

Phone: _____ E-mail: _____

Address: _____
Street   City   State   Zip

Your K-9 Math and Science Classroom Activities Resource

The AIMS Magazine is your source for standards-based, hands-on math and science investigations. Each issue is filled with teacher-friendly, ready-to-use activities that engage students in meaningful learning.

- *Four issues each year (fall, winter, spring, and summer).*

Current issue is shipped with all past issues within that volume.

| 1825 | Volume | XXV | 2010-2011 | $19.95 |
| 1826 | Volume | XXVI | 2011-2012 | $19.95 |

Two-Volume Combinations
| M21012 | Volumes | XXV & XXVI | 2010-12 | $34.95 |
| M21113 | Volumes | XXVI & XXVII | 2011-13 | $34.95 |

Complete volumes available for purchase:

| 1823 | Volume | XXIII | 2008-2009 | $19.95 |
| 1824 | Volume | XXIV | 2009-2010 | $19.95 |

AIMS Online—www.aimsedu.org

To see all that AIMS has to offer, check us out on the Internet at www.aimsedu.org. At our website you can preview and purchase AIMS books and individual activities, learn about State-Specific Science and Essential Math, explore professional development workshops and online learning opportunities, search our activities database, buy manipulatives and other classroom resources, and download free resources including articles, puzzles, and sample AIMS activities.

AIMS E-mail Specials
While visiting the AIMS website, sign up for our FREE e-mail newsletter with monthly subscriber-only specials. You'll also receive advance notice of new products.

Sign up today!

Subscribe to the AIMS Magazine

$19.95 a year!

AIMS Magazine is published four times a year.

Subscriptions ordered at any time will receive all issues for that year.

Call **1.888.733.2467** or go to www.aimsedu.org

AIMS Program Publications

Actions With Fractions, 4-9
The Amazing Circle, 4-9
Awesome Addition and Super Subtraction, 2-3
Bats Incredible! 2-4
Brick Layers II, 4-9
The Budding Botanist, 3-6
Chemistry Matters, 5-7
Counting on Coins, K-2
Cycles of Knowing and Growing, 1-3
Crazy About Cotton, 3-7
Critters, 2-5
Earth Book, 6-9
Electrical Connections, 4-9
Energy Explorations: Sound, Light, and Heat, 3-5
Exploring Environments, K-6
Fabulous Fractions, 3-6
Fall Into Math and Science*, K-1
Field Detectives, 3-6
Finding Your Bearings, 4-9
Floaters and Sinkers, 5-9
From Head to Toe, 5-9
Glide Into Winter With Math and Science*, K-1
Gravity Rules! 5-12
Hardhatting in a Geo-World, 3-5
Historical Connections in Mathematics, Vol. I, 5-9
Historical Connections in Mathematics, Vol. II, 5-9
Historical Connections in Mathematics, Vol. III, 5-9
It's About Time, K-2
It Must Be A Bird, Pre-K-2
Jaw Breakers and Heart Thumpers, 3-5
Looking at Geometry, 6-9
Looking at Lines, 6-9
Machine Shop, 5-9
Magnificent Microworld Adventures, 6-9
Marvelous Multiplication and Dazzling Division, 4-5
Math + Science, A Solution, 5-9
Mathematicians are People, Too
Mathematicians are People, Too, Vol. II
Mostly Magnets, 3-6
Movie Math Mania, 6-9
Multiplication the Algebra Way, 6-8
Out of This World, 4-8
Paper Square Geometry:
    The Mathematics of Origami, 5-12
Puzzle Play, 4-8
Popping With Power, 3-5

Positive vs. Negative, 6-9
Primarily Bears*, K-6
Primarily Earth, K-3
Primarily Magnets, K-2
Primarily Physics: Investigations in Sound, Light,
    and Heat Energy, K-2
Primarily Plants, K-3
Primarily Weather, K-3
Problem Solving: Just for the Fun of It! 4-9
Problem Solving: Just for the Fun of It! Book Two, 4-9
Proportional Reasoning, 6-9
Ray's Reflections, 4-8
Sensational Springtime, K-2
Sense-able Science, K-1
Shapes, Solids, and More: Concepts in Geometry, 2-3
The Sky's the Limit, 5-9
Soap Films and Bubbles, 4-9
Solve It! K-1: Problem-Solving Strategies, K-1
Solve It! 2nd: Problem-Solving Strategies, 2
Solve It! 3rd: Problem-Solving Strategies, 3
Solve It! 4th: Problem-Solving Strategies, 4
Solve It! 5th: Problem-Solving Strategies, 5
Solving Equations: A Conceptual Approach, 6-9
Spatial Visualization, 4-9
Spills and Ripples, 5-12
Spring Into Math and Science*, K-1
Statistics and Probability, 6-9
Through the Eyes of the Explorers, 5-9
Under Construction, K-2
Water, Precious Water, 4-6
Weather Sense: Temperature, Air Pressure, and Wind, 4-5
Weather Sense: Moisture, 4-5
What's Next, Volume 1, 4-12
What's Next, Volume 2, 4-12
What's Next, Volume 3, 4-12
Winter Wonders, K-2

Essential Math
Area Formulas for Parallelograms, Triangles, and Trapezoids, 6-8
Circumference and Area of Circles, 5-7
Effects of Changing Lengths, 6-8
Measurement of Prisms, Pyramids, Cylinders, and Cones, 6-8
Measurement of Rectangular Solids, 5-7
Perimeter and Area of Rectangles, 4-6
The Pythagorean Relationship, 6-8

Spanish Edition
Constructores II: Ingeniería Creativa Con Construcciones
    LEGO®, 4-9
    The entire book is written in Spanish. English pages not included.

* Spanish supplements are available for these books. They are only available as downloads from the AIMS website. The supplements contain only the student pages in Spanish; you will need the English version of the book for the teacher's text.

For further information, contact:
AIMS Education Foundation • P.O. Box 8120 • Fresno, California 93747-8120
www.aimsedu.org • 559.255.6396 (fax) • 888.733.2467 (toll free)

Duplication Rights

No part of any AIMS books, magazines, activities, or content—digital or otherwise—may be reproduced or transmitted in any form or by any means—including photocopying, taping, or information storage/retrieval systems—except as noted below.

Standard Duplication Rights

- A person or school purchasing AIMS activities (in books, magazines, or in digital form) is hereby granted permission to make up to 200 copies of any portion of those activities, provided these copies will be used for educational purposes and only at one school site.
- Workshop or conference presenters may make one copy of any portion of a purchased activity for each participant, with a limit of five activities per workshop or conference session.
- All copies must bear the AIMS Education Foundation copyright information.

Standard duplication rights apply to activities received at workshops, free sample activities provided by AIMS, and activities received by conference participants.

Unlimited Duplication Rights

Unlimited duplication rights may be purchased in cases where AIMS users wish to:
- make more than 200 copies of a book/magazine/activity,
- use a book/magazine/activity at more than one school site, or
- make an activity available on the Internet (see below).

These rights permit unlimited duplication of purchased books, magazines, and/or activities (including revisions) for use at a given school site.

Activities received at workshops are eligible for upgrade from standard to unlimited duplication rights.

Free sample activities and activities received as a conference participant are not eligible for upgrade from standard to unlimited duplication rights.

State-Specific Science modules are licensed to one classroom/one teacher and are therefore not eligible for upgrade from standard to unlimited duplication rights.

Upgrade Fees

The fees for upgrading from standard to unlimited duplication rights are:
- $5 per activity per site,
- $25 per book per site, and
- $10 per magazine issue per site.

The cost of upgrading is shown in the following examples:
- activity: 5 activities x 5 sites x $5 = $125
- book: 10 books x 5 sites x $25 = $1250
- magazine issue: 1 issue x 5 sites x $10 = $50

Purchasing Unlimited Duplication Rights

To purchase unlimited duplication rights, please provide us the following:
1. The name of the individual responsible for coordinating the purchase of duplication rights.
2. The title of each book, activity, and magazine issue to be covered.
3. The number of school sites and name of each site for which rights are being purchased.
4. Payment (check, purchase order, credit card)

Requested duplication rights are automatically authorized with payment. The individual responsible for coordinating the purchase of duplication rights will be sent a certificate verifying the purchase.

Internet Use

AIMS materials may be made available on the Internet if all of the following stipulations are met:
1. The materials to be put online are purchased as PDF files from AIMS (i.e., no scanned copies).
2. Unlimited duplication rights are purchased for all materials to be put online for each school at which they will be used. (See above.)
3. The materials are made available via a secure, password-protected system that can only be accessed by employees at schools for which duplication rights have been purchased.

AIMS materials may not be made available on any publicly accessible Internet site.